SELECTED WORKS OF GIUSEPPE PEANO

Selected works of
GIUSEPPE PEANO

Translated and edited, with a
biographical sketch and bibliography, by
HUBERT C. KENNEDY

University of Toronto Press

© University of Toronto Press 1973

Toronto and Buffalo

Printed in the United States of America

ISBN 0-8020-5267-3

Microfiche ISBN 0-8020-0161-0

LC 70-185719

AMS 1970 Subject classification 01A75

Contents

Preface

The influence of Giuseppe Peano on turn-of-the-century mathematics and logic was great. In the decade preceding 1900, for example, Peano and members of his school were leaders in contemporary developments in logic. If the initiative then passed to Bertrand Russell in England, this was because Peano inspired him when they met at the International Philosophical Congress in Paris in August 1900. Russell has said (in his *Autobiography*): 'The Congress was a turning point in my intellectual development, because I there met Peano.'

Peano's influence on mathematics is only partly illustrated by what is named after him, for example the postulates for the natural numbers and the space-filling curve. His influence was perhaps greatest, as he himself said in 1915, in the development of the calculus. He seemed particularly skilled in the discovery of counterexamples to currently accepted notions and theorems. The space-filling curve was a spectacular example, but there were many others. His call for rigour in the treatment of notions such as surface area was not new, but his contributions played an influential role.

In view of the importance of Peano's work, it is surprising that no complete article by him has previously appeared in English. Indeed, until recently it was difficult for anyone interested in reading the original articles of Peano to find them. This need has largely been met by the publication, in three volumes, of the *Opere scelte di Giuseppe Peano* (Rome: Edizioni Cremonese, 1957–9) and the facsimile reproduction of the *Formulario mathematico* (Rome: Edizioni Cremonese, 1960). The problem of language remains, however, especially for English-speaking readers. In the first place, few English-speaking mathematicians and logicians read Italian (the language most used by Peano). Secondly, although the difficulty of reading *Latino sine flexione* (or Latin-without-grammar, Peano's own invention) is more apparent than real, nonetheless few readers get over the initial shock of its strange appearance. The present translations should help answer these difficulties.

My interest in Peano was aroused when, as a student at the University of Milan during the academic year 1957–8, I met Professors Ugo Cassina and Ludovico Geymonat, both of whom had studied with Peano. (I attended the lectures of Professor Geymonat in the philosophy of mathematics; Professor Cassina directed my reading of the logical works of Peano.) Professor Cassina was then preparing the three-volume *Opere scelte di Giuseppe Peano*, and I was particularly impressed by his devotion to his former teacher. Several years later, on 29 October 1963, I wrote to him, asking whether he planned to write a biography of Peano and offering my assistance. He suggested that we collaborate, meaning by this that I was to write the biography in English using the material that he had collected, but before this project could get underway, Professor Cassina died unexpectedly on 5 October 1964. I expected a sabbatical leave for the academic year 1966–7 and determined to spend my leave in Italy collecting material for the biography. In the meantime, partly as preparation for my trip to Italy, I translated for my own use several articles of Peano. I was struck by the fact that although many of Peano's articles had already been translated into one or more languages, no complete work of Peano had previously been published in English. Here was a project that should precede the writing of the biography. That is the origin of the present volume.

In making the selection of articles to be translated, I have had in mind the interests of mathematicians and logicians (and teachers of these subjects). Roughly, two types of articles have been included: those that the reader may be expected to know about already, and want to read for himself (e.g., the space-filling curve); and those I believe should also be read in order to have a balanced picture of Peano's achievements (e.g., those which illustrate Peano's own judgment that he had most contributed to the development of the calculus). I have not included articles that were peripheral to Peano's interests (e.g., in actuarial mathematics) or that could be expected to be peripheral to the interests of mathematicians and logicians (e.g., propaganda articles for *Latino sine flexione*, which for that matter are untranslatable). Some attempt has been made to include selections from different periods of Peano's career. Like many mathematicians, he did his most creative work early, but his mature writings are valuable, especially from a pedagogical point of view.

Considering the influence of Peano on the development of mathematical symbolism, I have tried to retain his original notations. For typographical reasons, however, this has not always been possible. Something more should be said about the problems that arise in translating Peano. His prose style has most often been described as 'sparse' (*scarno*). It was certainly never literary and often it was not good Italian. Making a complete paragraph of, for example, 'etc.' is neither good English nor good Italian. On the whole his prose is clear and direct. I have tried to retain this aspect of his style, and I have no doubt erred on the side of a too closely literal translation. Of the twenty-three works selected for translation, fifteen were originally in Italian, five in French, one in scholastic Latin, and two in *Latino sine flexione*. The last poses special problems for the translator since it is by its nature non-grammatical. I have tried to translate these articles in the style of Peano's current use of Italian, which, perhaps being influenced by his dedication to *Latino sine flexione*, tended to become ever more sparse and simple.

I am grateful to all who inspired, encouraged, and supported me during the preparation of these translations: Providence College for a year's leave of absence in Italy; the Department of State for financial and other support, under the Fulbright program, during my two visits to Italy; Brown University for unlimited access to its rich libraries; the University of Turin and its professors for access to the mathematics library and a warm reception; the late Ugo Cassina, Peano's three charming nieces, and Professors L. Geymonat and E. Carruccio for stimulating conversation; and Professors C.B. Boyer and K.O. May for their continued encouragement. I especially wish to thank Miss L. Ourom, Editor (Science) of the University of Toronto Press, for her assistance in the preparation of the manuscript. The changes she suggested were always improvements. I also wish to thank the Unione matematica italiana for its kind permission to publish the translations of articles which appear in *Opere scelte* and Peano's three nieces, Carola, Caterina, and Maria Peano, for permission to publish the translations of the remaining three items (chapters v; ix; section 2; and xvi).

This work has been published with the aid of the Publications Fund of the University of Toronto Press.

Except where indicated, all translations are mine, and all insertions, comments, and footnotes enclosed in square brackets are mine.

Providence, Rhode Island   H.C.K.
January 1971

LIFE AND WORKS

Biographical sketch of Giuseppe Peano

Giuseppe Peano was born on 27 August 1858, the second of the five children of Bartolomeo and Rosa Cavallo. His brother Michele was seven years older. There were to be two younger brothers, Francesco and Bartolomeo, and a sister, Rosa. Peano's first home was the farm 'Tetti Galant,' near the village of Spinetta, three miles from Cuneo and fifty miles south of Turin. When Peano entered school, both he and his brother walked the distance to Cuneo each day, but the family later moved to Cuneo so that the children would not have so far to go. The older brother became a successful accountant and remained in Cuneo. In 1967 Tetti Galant was still in the possession of the Peano family.

Peano's maternal uncle, Michele Cavallo, a priest and lawyer, lived in Turin. On his invitation, Peano moved to that city when he was twelve or thirteen years old. There he received private lessons, some from his uncle, and studied on his own, so that he was able to pass the lower secondary exam of the Cavour School in 1873. He then attended as a regular pupil, completing the upper secondary program in 1876. His performance won him a room-and-board scholarship at the Collegio delle Provincie, which was established to assist students from the provinces to attend the university. Peano duly enrolled at the university in the fall of 1876.

Peano's professors of mathematics at the University of Turin included Enrico D'Ovidio, Angelo Genocchi, Francesco Siacci, Giuseppe Basso, Francesco Faà di Bruno, and Giuseppe Erba. On 16 July 1880 he completed his final exam 'with high honours.' For the academic year 1880–81 he was assistant to Enrico D'Ovidio, and from the fall of 1881 he was assistant to and later substitute for Angelo Genocchi, until Genocchi's death in 1889.

On 21 July 1887 Peano married Carola Crosio, daughter of Luigi Crosio, a genre painter, especially of Pompeian and seventeenth-century scenes. She was interested in opera and they often went to performances at the Royal Theatre of Turin, where the first performances of *Manon*

Lescaut and *Bohème* of Puccini (who was almost exactly Peano's age) were given in 1893 and 1896, respectively.

On 1 December 1890, after regular competition, Peano was named Extraordinary Professor of Infinitesimal Calculus at the University of Turin. He was promoted to Ordinary Professor in 1895. He had already, in 1886, been named Professor at the Military Academy, which was situated close to the university. He gave up his position at the Military Academy in 1901, but retained his position of Professor at the university until his death in 1932, transferring in 1931 to the Chair of Complementary Mathematics. He was elected to membership in a number of scientific societies, among them the Academy of Sciences of Turin, in which he played a very active role. He was also made a knight of the Crown of Italy and of the Order of Saints Maurizio and Lazzaro. Although he was not active politically, his views tended toward socialism, and he once invited a group of textile workers, who were on strike, to a party at his home. During the First World War he advocated a closer federation of the allied countries, to better prosecute the war and, after the peace, to form the nucleus of a world federation. In religion he was Catholic, but non-practising.

Peano's father died in 1888; his mother, in 1910. Although he was rather frail as a child, Peano's health was generally good, his most serious illness being an attack of smallpox in August 1889. Peano died of a heart attack the morning of 20 April 1932, after having taught his regular class the previous afternoon. At his request the funeral was very simple, and he was buried in the Turin General Cemetery. Peano was survived by his wife, his sister, and a brother. He had no children. In 1963 his remains were transferred to the family tomb in Spinetta.

Peano is perhaps most widely known as a pioneer of symbolic logic and a promoter of the axiomatic method, but he considered his work in analysis most important. In 1915 he printed a list of his publications, adding: 'My works refer especially to infinitesimal calculus, and they have not been entirely useless, seeing that, in the judgment of competent persons, they contributed to the constitution of this science as we have it today.' This 'judgment of competent persons' refers in part to the *Encyklopädie der mathematischen Wissenschaften*, in which Pringsheim lists two of Peano's books among the nineteen most important calculus texts since the time of Euler and Cauchy. The first of these was Peano's first major publication and is something of an oddity in the history of mathematics, since the title page gives the author as Angelo Genocchi, not Peano. The title is 'Differential Calculus and Fundamentals of Integral Calculus' and the title page states that it is 'published with additions by Dr Giuseppe Peano.' An explanation of the origin of this book is given in a letter that Peano wrote Genocchi on 7 June 1883:

Esteemed professor,

A few days ago I was at Bocca book-publishers and the director, who is named Lerda, I believe, showed a great desire to publish part of a calculus text during the summer vacation, either written by you or according to your method of teaching: there is no need for me to add how useful such a work would be.

Would you please be so kind as to let me know whether it would be possible to firm up this matter in some way, i.e., if you do not wish to publish the text yourself, perhaps you might think it possible for me to write it, following your lessons, and if I did that, would you be willing to examine my manuscript before publication, or at any rate give me your valued suggestions and look over the proofs of the fascicles as they come off the press; or indeed, in the case that you do not wish to publish this yourself, it might not displease you if I went ahead and published the text, saying that it was *compiled according to your lessons*, or at least citing your name in the Preface, because a great part of the treatment of the material would be yours, inasmuch as I learned it from you.

Please allow me, learned professor, to make a point of saying that I will do my best to see that all comes out well, and believe me

your most devoted student

G. Peano
Turin, 7 June 1883

Genocchi's reception of this letter is expressed in a letter sent to his friend Judge P. Agnelli two months later. In part, he says: 'I could make a complete calculus course with my lessons, but I don't feel up to writing it ... and seeing that Dr Peano, my assistant and substitute, and former student, is willing to take on the trouble, I thought it best not to oppose the project and let him handle it as best he can.'

The publishing firm of Bocca Brothers announced publication for October 1883 of the first volume of Genocchi's *Corso di calcolo infinitesimale* (Course in Infinitesimal Calculus), edited by Peano. When the publication did not appear on time, Felice Casorati wrote to Peano asking about it. A very friendly exchange of letters followed. Peano replied that about 100 pages had been printed and were already being used by his students. The complete book, however, did not appear until the fall of 1884. Before the end of that year Genocchi had had published in mathematical journals of Italy, France, and Belgium the following declaration:

Recently the publishing firm of Bocca Brothers published a volume entitled *Calcolo differenziale e principii di calcolo integrale* (Differential Calculus and Fundamentals of Integral Calculus). At the top of the title page is placed my name, and in the Preface it is stated that besides the course given by me at the University of Turin the volume contains important additions, some modifications, and various annotations, which are placed first. So that nothing will be attributed to me which is not mine, I must declare that I have had no part in the compilation of the aforementioned book and that everything is due to that outstanding young man Dr Giuseppe Peano, whose name is signed to the Preface and Annotations.

Angelo Genocchi

It is easy to imagine that, on reading this declaration, many people thought that Peano was using the name of Genocchi (who was fairly well known in Europe at that time) just to get his own research published, or at least was guilty of bad faith in dealing with his former teacher. Indeed, Charles Hermite, who had written to Genocchi on 6 October 1884 congratulating him on its publication, wrote again on 31 October to commiserate with Genocchi on having such an indiscreet and unfaithful assistant. Genocchi complained to his friend Placido Tardy, who wrote to Genocchi on 8 November 1884: 'After your letter, I sent a visiting card as a thank you note to Peano. I am surprised by what you tell me about the manner in which he has conducted himself toward you, in publishing your lessons.' But it seems that Genocchi, who had a reputation for being quick-tempered, soon recovered and three weeks later Tardy could write: 'I, too, am persuaded that Peano never suspected that he might be lacking in regard toward you.'

In the Preface to his second book, *Applicazioni geometriche del calcolo infinitesimale* (Geometrical Applications of the Infinitesimal Calculus), Peano explicitly states that the first book was authorized by Genocchi, but that he himself assumed full responsibility for it. After Genocchi's death, Peano wrote a necrology of Genocchi for publication in the university yearbook, in which he reaffirms this, and justifies the changes made by the approval the book obtained from critics.

Of the many notable things in this book, the *Encyklopädie der mathematischen Wissenschaften* cites: theorems and remarks on limits of indeterminate expressions, and the pointing out of errors in the better texts then in use; a generalization of the mean-value theorem for derivatives; a theorem on uniform continuity of functions of several variables; theorems on the existence and differentiability of implicit functions; an example of a function whose partial derivatives do not commute; conditions for expressing a function of several variables by a Taylor's formula; a counterexample to the current theory of minima; and rules for integrating rational functions when roots of the denominator are not known.

The other text of Peano cited in the *Encyklopädie* was the two-volume *Lezioni di analisi infinitesimale* (Lessons in Infinitesimal Analysis) of 1893. This work contains fewer new results, but is notable for its rigour and clarity of exposition.

Peano began his scientific publication in 1881–82 with four articles dealing with geometry and algebraic forms. They were along the lines of work done by his professors E. D'Ovidio and F. Faà di Bruno. His work in analysis began in 1883 with 'On the integrability of functions' (5). (The number in parentheses refers to the Chronological List, where the original title of the article is given.) This note contains original notions of integrals and areas. Peano was the first to show that the first-order differential equation $y' = f(x, y)$ is soluble on the sole assumption that f is continuous. His first proof (9) dates from 1886, but its rigour leaves something to be desired. This result was generalized in 1890 to systems of differential equations (27) using a different method of proof. This work is notable, also, for containing the first explicit statement of the axiom of choice. Peano rejected it as being outside the ordinary logic used in mathematical proofs.

Already in the 'Calculus' of 1884 Peano had given many counterexamples to commonly accepted notions in mathematics, but his most famous example was the space-filling curve (24) published in 1890. This curve is given by continuous parametric functions and goes through every point in a square as the parameter ranges over some interval. It can be defined as the limit of a sequence of more 'ordinary' curves. Peano was so proud of this discovery that he had one of the curves in the sequence put on the terrace of his home, in black tiles on white. Some of Peano's work in analysis was quite original and he has not always been given credit for his priority, but much of his publication was designed to clarify definitions and theories then current and make them more rigorous.

Peano's work in logic and the foundations of mathematics may be considered together, although he never subscribed to Bertrand Russell's reduction of mathematics to logic. Peano's first publication in logic was a twenty-page preliminary section, 'Operations of deductive logic,' in his book of 1888, *Calcolo geometrico secondo l'Ausdehnungslehre di H. Grassmann* (Geometrical Calculus according to the *Ausdehnungslehre* of H. Grassmann). This section has almost no connection with what follows it. It is a synthesis of, and improvement on, some of the work of Boole, Schröder, Peirce, and MacColl. But by the following year, with the publication of *Arithmetices principia, nova methodo exposita* (The Principles of Arithmetic, Presented by a New Method) (16), Peano has not only improved his logical symbolism, but uses his new method to achieve important new results in mathematics, for this short booklet contains Peano's first statement of his famous postulates for the natural numbers, perhaps the best known of all his creations. His research was done independently of the work of Dedekind, who had the previous year published an analysis of the natural numbers which was essentially that of Peano, without, however, the same clarity. Peano's work also made important innovations in logical notation: ε for set membership, and a new notation for universal quantification. Peano's influence on the development of logical notation was great. Jean van Heijenoort has said: 'The ease with which we read Peano's booklet today shows how much of his notation has found its way, either directly or in a somewhat modified form, into contemporary logic.' (*From Frege to Gödel, a Source Book in Mathematical Logic, 1879–1931,* [Cambridge, Mass., 1967], p. 84. This book contains a translation of part of Peano's 'Principles of Arithmetic.')

Peano was less interested in logic itself than in logic used in mathematics. For this reason he referred to his system as 'mathematical logic.' This may partly explain his failure to develop rules of inference. Some of Peano's omissions are quite remarkable, but his influence on the development of logic was tremendous. Hans Freudenthal said of the Paris Philosophical Congress of 1900: 'In the field of the philosophy of science the Italian phalanx was supreme: Peano, Burali-Forti, Padoa, Pieri absolutely dominated the discussion. For Russell, who read a paper that was philosophical in the worst sense, Paris was the road to Damascus.' ('The Main Trends in the Foundations of Geometry in the 19th Century,' *Logic, Methodology and Philosophy of Science* [Stanford, 1962], p. 616.) Indeed, Russell has remarked that the two men who most influenced his philosophical development were G.E. Moore and Peano.

In 1891 Peano founded the journal *Rivista di matematica*, in which the results of his research, and that of his followers, in logic and the foundations of mathematics were published. It was in this journal, in 1892, that he announced the Formulario project, a project which was to take much of his mathematical and editorial energies for the next sixteen years. The end result of this project would be, he hoped, the publication of a collection of all known theorems in the various branches of mathematics. The notations of his mathematical logic were to be used and proofs of the theorems were to be given. There were five editions of the *Formulario*: the first appeared in 1895, and the last, completed in 1908, contained some 4200 theorems.

In addition to his research in logic and arithmetic, Peano also applied the axiomatic method to other fields, notably geometry, for which he gave several systems of axioms.

Of more importance in geometry was Peano's popularization of the vectorial methods of H. Grassmann, beginning with the publication, already mentioned, in 1888 of the 'Geometrical Calculus according to the *Ausdehnungslehre* of H. Grassmann.' Grassmann's own publications have been criticized for their abstruseness. Nothing could be clearer than Peano's presentation, and the impetus he gave to the Italian school of vector analysis was great.

Enough has been said to show that Peano was interested in, and left his mark on, a number of fields of mathematics. We are making no attempt here to cover everything, but perhaps his work in numerical calculation should be mentioned. Peano devoted a good deal of his effort to this, especially to generalizing Taylor's formula and finding new forms for the remainder. Much of this work has been outmoded by the advent of the high-speed computer, though even in this field something of Peano's influence remains.

The year 1903 saw the beginning of a shift in Peano's interest, from mathematics to the promotion of an international auxiliary language, this shift being nearly completed in 1908 with his election to the presidency of the old Volapük Academy, later renamed 'Academia pro Interlingua.' For the rest of his life Peano was an ardent apostle of the idea of an international auxiliary language. His own invention, *Latino sine flexione* or Latin without grammar, was very similar to that later adopted by the International Auxiliary Language Association and called 'Interlingua.' Peano was an important figure in the history of the artificial language movement. The publication in 1915 of his *Vocabulario commune ad latino-italiano-français-english-deutsch* (Vocabulary Common to Latin-Italian-French-English-German) was a prime event.

It has been said that the apostle in Peano impeded the work of the mathematician. This is no doubt true, especially of his later years, but there can be no question of his very real influence on the development of mathematics. He contributed in no small part to the popularity of the axiomatic method, and his discovery of the space-filling curve must be considered remarkable. Although many of his notions, such as area and integral, were 'in the air,' his originality cannot be denied. He was not an imposing person and his gruff voice with its high degree of lalation (the pronunciation of 'r' so that it sounds like 'l') could hardly have been attractive, but his gentle personality commanded respect, and his keen intellect inspired disciples to the end. His most devoted disciple was one of his last, Ugo Cassina, who shared his interest in mathematics and his zeal for Interlingua. Peano was very generous in promoting the work of young mathematicians. During the First World War, when the Academy of Sciences of Turin limited the number of pages of their Proceedings, Peano, who habitually offered for publication the papers of his school-teacher friends, was most affected. Much of Peano's mathematics is now only of historical interest, but his call for clarity and rigour in mathematics and its teaching remains relevant today – and few have expressed this call more forcefully.

Chronological list
of the publications of Giuseppe Peano

In the chronological list published in the three-volume *Opere scelte*, Ugo Cassina used the consecutive numbering of Peano's publications which had already been used by Peano himself. He also distinguished between mathematical and philological publications, leaving only one publication outside these two categories (and outside the consecutive numbering). Since the *Opere scelte* will be the most available source, for most readers, of the original articles, I have kept the numbering of Cassina and have used a decimal numbering to insert new titles into his list (increasing the number of titles by almost twenty per cent). I have also followed Cassina's use of Roman numerals to distinguish monographs and volumes (of at least 64 pages), primes to denote translations and reprints, and the asterisk to signalize works pertaining to philology. Thus *160'-17' is a reprint of the 160th work and the 17th work in philology. Titles of journals are those of the *Union List of Serials*; abbreviations of journal titles are those of *Mathematical Reviews* (with the addition of A.p.I. and R.d.M.). In cases where the original article was untitled, I have given a brief descriptive phrase in English. This suffices to show that the article was untitled since none of Peano's publications were in English, nor has a translation of a complete article into English been previously published.

The following abbreviations of journal titles are used:

A.p.I.   Academia pro Interlingua
Amer. J. Math.   American Journal of Mathematics
Ann. mat. pura appl.   Annali di matematica pura ed applicata
Atti Accad. naz. Lincei, Rend., Cl. sci. fis. mat. nat.   Atti della Accademia Nazionale dei Lincei, Rendiconti, Classe di scienze fisiche, matematiche e naturali
Atti Accad. sci. Torino, Cl. sci. fis. mat. nat.   Atti della Accademia delle scienze di Torino, Classe di scienze fisiche, matematiche e naturali
Boll. Un. mat. ital.   Bollettino della Unione matematica italiana
Enseignement math.   L'Enseignement mathématique

Giorn. mat. Battaglini Giornale di matematiche di Battaglini
Giorn. mat. finanz. Giornale di matematica finanziaria: Rivista tecnica
del credito e della previdenza
Math. Ann. Mathematische Annalen
Mathesis Mathesis: Recueil mathématique à l'usage des écoles spéciales
et des établissements d'instruction moyenne
Monatsh. Math. Monatshefte für Mathematik
Period. mat. Periodico di matematiche
R.d.M. Rivista di matematica (Revue de mathématiques)
Rend. Circ. mat. Palermo Rendiconti del Circolo matematico di Palermo
Scientia Scientia: International Review of Scientific Synthesis
Wiadom. mat. Rocziki Polskiego towarzystwa matematycznego, ser. II:
Wiadomości matematyczne

1881

1 'Costruzione dei connessi (1, 2) e (2, 2),' *Atti Accad. sci. Torino*, 16
(1880–1), 497–503

2 'Un teorema sulle forme multiple,' *Atti Accad. sci. Torino*, 17
(1881–2), 73–9

3 'Formazioni invariantive delle corrispondenze,' *Giorn. mat.
Battaglini*, 20 (1881), 79–100

1882

4 'Sui sistemi di forme binarie di egual grado, e sistema completo di
quante si vogliano cubiche,' *Atti Accad. sci. Torino*, 17 (1881–2),
580–6

1883

5 'Sull'integrabilità delle funzioni,' *Atti Accad. sci. Torino*, 18
(1882–3), 439–46

6 'Sulle funzioni interpolari,' *Atti Accad. sci. Torino*, 18 (1882–3),
573–80

1884

7 Two letters to the Editor, *Nouvelles annales de mathématiques*, (3)
3 (1884), 45–7, 252–6

I, 8 Angelo Genocchi, *Calcolo differenziale e principii di calcolo
integrale, pubblicato con aggiunte dal Dr. Giuseppe Peano* (Torino:
Bocca, 1884), pp. xxxii + 338

I', 8' Angelo Genocchi, *Differentialrechnung und Grundzüge der
Integralrechnung, herausgegeben von Giuseppe Peano*, trans. by

G. Bohlmann and A. Schepp, with a preface by A. Mayer
(Leipzig: Teubner, 1899), pp. viii + 399

1886

9 'Sull'integrabilità delle equazioni differenziali del primo ordine,'
Atti Accad. sci. Torino, 21 (1885–6), 677–85

1887

10 'Integrazione per serie delle equazioni differenziali lineari,' *Atti
Accad. sci. Torino*, 22 (1886–7), 437–46

II, 11 *Applicazioni geometriche del calcolo infinitesimale* (Torino:
Bocca, 1887), pp. xii + 336

1888

12 'Intégration par séries des équations différentielles linéaires,'
Math. Ann., 32 (1888), 450–6

13 'Definizione geometrica delle funzioni ellittiche,' *Giorn. mat.
Battaglini*, 26 (1888), 255–6

13′ 'Definição geometrica das funcções ellipticas,' *Jornal de sciencias
mathematicas e astronomicas*, 9 (1889), 24–5

III, 14 *Calcolo geometrico secondo l'Ausdehnungslehre di H. Grassmann,
preceduto dalle operazioni della logica deduttiva* (Torino: Bocca,
1888), pp. xii + 170

14.1 Communications to the Circolo (on a note of F. Giudice and on
his response), *Rend. Circ. mat. Palermo*, 2 (1888), 94, 187–8

15 'Teoremi su massimi e minimi geometrici, e su normali a curve
e superficie,' *Rend. Circ. mat. Palermo*, 2 (1888), 189–92

1889

16 *Arithmetices principia, nova methodo exposita* (Torino: Bocca,
1889), pp. xvi + 20

17 'Sur les wronskiens,' *Mathesis*, 9 (1889), 75–6, 110–12

18 *I principii di geometria logicamente esposti* (Torino: Bocca, 1889),
pp. 40

19 'Une nouvelle forme du reste dans la formule de Taylor,'
Mathesis, 9 (1889), 182–3

20 'Su d'una proposizione riferentesi ai determinanti jacobiani,'
Giorn. mat. Battaglini, 27 (1889), 226–8

21 'Angelo Genocchi,' *Annuario R. Università di Torino* (1889–90),
195–202

40 Open letter to Prof. G. Veronese, *R.d.M.*, 1 (1891), 267–9

41 'Il teorema fondamentale di trigonometria sferica,' *R.d.M.*, 1 (1891), 269

42 'Sulla formula di Taylor,' *Atti Accad. sci. Torino*, 27 (1891–2), 40–6

42′ 'Ueber die Taylor'sche Formel,' Anhang III in (I′, 8′), 359–65

42.1 'Questions proposées, no. 1599,' *Nouvelles annales de mathématiques*, (3) 10 (1891), 2*

a *Rivista di matematica*, vol. 1 (Torino: Bocca, 1891), pp. iv + 272

1892

43 Comments on an unsolved problem, *R.d.M.*, 2 (1892), 1–2

44 'Sommario del libro x d'Euclide,' *R.d.M.*, 2 (1892), 7–11

44.1 (With F. Giudice) Review of Domenico Amanzio, *Elementi di algebra elementare*, *R.d.M.*, 2 (1892), 14–17

45 'Sur la définition de la dérivée,' *Mathesis*, (2) 2 (1892), 12–14

46 'Osservazioni sul "Traité d'analyse par H. Laurent," ' *R.d.M.*, 2 (1892), 31–4

46.1 (Unsigned) 'Enrico Novarese,' *R.d.M.*, 2 (1892), 35

47 'Esempi di funzioni sempre crescenti e discontinue in ogni intervallo,' *R.d.M.*, 2 (1892), 41–2

48 Proposed question no. VI, *R.d.M.*, 2 (1892), 42

49 'Sur le théorème général relatif à l'existence des intégrales des équations différentielles ordinaires,' *Nouvelles annales de mathématiques*, (3) 11 (1892), 79–82

50 'Generalizzazione della formula di Simpson,' *Atti Accad. sci. Torino*, 27 (1891–2), 608–12

51 'Dimostrazione dell'impossibilità di segmenti infinitesimi costanti,' *R.d.M.*, 2 (1892), 58–62

52 'Sulla definizione del limite d'una funzione,' *R.d.M.*, 2 (1892), 77–9

53 Review of Albino Nagy, *Lo stato attuale ed i progressi della logica*, *R.d.M.*, 2 (1892), 80

54 A brief reply to Prof. Veronese, *Rend. Circ. mat. Palermo*, 6 (1892), 160

55 Letter to the Editor (Extrait d'une lettre de M. Peano à M. Brisse), *Nouvelles annales de mathématiques*, (3) 11 (1892), 289

56 Review of G. Veronese, *Fondamenti di geometria a più dimensioni*, etc., *R.d.M.*, 2 (1892), 143–4

b *Rivista di matematica*, vol. 2 (Torino: Bocca, 1892), pp. iv + 215.

70.3   (With G. Vivanti) *Teoria dei gruppi di punti* (Torino: Fodratti e
Lecco, 1894)

   d   *Rivista di matematica*, vol. 4 (Torino: Bocca, 1894), pp. iv + 198

1895

v, 71   *Formulaire de mathématiques*, vol. 1 (Torino: Bocca, 1895),
pp. vii + 144

72   (With G. Vailati) 'Logique mathématique,' (v, 71), 1–8

73   (With F. Castellano) 'Opérations algébriques,' (v, 71), 8–22

74   (With C. Burali-Forti) 'Arithmétique,' (v, 71), 22–8

75   'Classes de nombres,' (v, 71), 58–65

76   Review of F. Castellano, *Lezioni di meccanica razionale, R.d.M.*, 5
(1895), 11–18

77   'Il principio delle aree e la storia di un gatto,' *R.d.M.*, 5 (1895),
31–2

78   'Sulla definizione di integrale,' *Ann. mat. pura appl.*, (2) 23 (1895),
153–7

78′   'Ueber die Definition des Integrals,' Anhang IV in (I′, 8′), 366–70

79   'Sopra lo spostamento del polo sulla terra,' *Atti Accad. sci.
Torino*, 30 (1894–5), 515–23

80   'Sul moto del polo terrestre,' *Atti Accad. sci. Torino*, 30 (1894–5),
845–52

81   Letter to the Editor, *Monatsh. Math.*, 6 (1895), 204

82   Review of G. Frege, *Grundgesetze der Arithmetik, begriffs-
schriftlich abgeleitet*, vol. 1, *R.d.M.*, 5 (1895), 122–8

83   'Elenco bibliografico sull' "Ausdehnungslehre" di
H. Grassmann,' *R.d.M.*, 5 (1895), 179–82

84   'Sul moto d'un sistema nel quale sussistono moti interni
variabili,' *Atti Accad. naz. Lincei, Rend., Cl. sci. fis. mat. nat.*, (5)
4-II (1895), 280–2

85   'Trasformazioni lineari dei vettori di un piano,' *Atti Accad. sci.
Torino*, 31 (1895–6), 157–66

86   (With E. D'Ovidio) 'Relazione sulla memoria del Prof.
Francesco Giudice, intitolata: "Sull'equazione del 5° grado," '
Atti Accad. sci. Torino, 31 (1895–6), 199

86.1   Reply to question no. 288, *L'intermédiaire des mathématiciens*, 2
(1895), 83

   e   *Rivista di matematica*, vol. 5 (Torino: Bocca, 1895), pp. iv + 195

1896

87   'Sul pendolo di lunghezza variabile,' *Rend. Circ. mat. Palermo*, 10
(1896), 36–7

88 'Introduction au tome II du *Formulaire de mathématiques*,' *R.d.M.*, 6 (1896–9), 1–4

89 'Sul moto del polo terrestre,' *Atti Accad. naz. Lincei, Rend.*, *Cl. sci. fis. mat. nat.*, (5) 5-I (1896), 163–8

90 'Saggio di calcolo geometrico,' *Atti Accad. sci. Torino*, 31 (1895–6), 952–75

90' *Zarys Rachunku geometrycznego*, trans. by S. Dickstein (Warsaw, 1897), pp. 28

90'' *Entwicklung der Grundbegriffe des geometrischen Calculs*, trans. by A. Lanner (Salzburg, 1897–8), pp. 24

90.1 Reply to question no. 60, *L'intermédiaire des mathématiciens*, 3 (1896), 39

90.2 Reply to question no. 101, *L'intermédiaire des mathématiciens*, 3 (1896), 61

90.3 Reply to question no. 362, *L'intermédiaire des mathématiciens*, 3 (1896), 69

90.4 Reply to question no. 80, *L'intermédiaire des mathématiciens*, 3 (1896), 87

90.5 Reply to question no. 719, *L'intermédiaire des mathématiciens*, 3 (1896), 169

1897

91 'Studii di logica matematica,' *Atti Accad. sci. Torino*, 32 (1896–7), 565–83

91' 'Ueber mathematische Logik,' Anhang I in (I', 8'), 336–52

92 'Sul determinante wronskiano,' *Atti Accad. naz. Lincei, Rend.*, *Cl. sci. fis. mat. nat.*, (5) 6-I (1897), 413–15

VI, 93 *Formulaire de mathématiques*, vol. 2, section 1 : *Logique mathématique* (Turin : Bocca, 1897), pp. 64

93' 'Logica matematica,' *Verhandlungen des ersten internationalen Mathematiker-Kongresses, in Zürich vom 9. bis 11. August 1897* (Leipzig, 1898), p. 299

94 'Generalità sulle equazioni differenziali ordinarie,' *Atti Accad. sci. Torino*, 33 (1897–8), 9–18

95 (With E. D'Ovidio and C. Segre) 'Relazione sulla memoria "I principii della geometria di posizione composti in sistema logico-deduttivo" del Prof. M. Pieri,' *Atti Accad. sci. Torino*, 33 (1897–8), 148–50

1898

96 'Sulle formule di logica,' *R.d.M.*, 6 (1896–99), 48–52

97 Reply to a letter of G. Frege, *R.d.M.*, 6 (1896–99), 60–1

98 'Analisi della teoria dei vettori,' *Atti Accad. sci. Torino*, 33 (1897–8), 513–34

VII, 99 *Formulaire de mathématiques*, vol. 2, section 2 (Turin: Bocca, 1898), pp. viii + 60

99' 'Definitionen der Arithmetik,' Anhang II in (I', 8'), 353–8

100 (With others) 'Additions et corrections à F₂,' *R.d.M.*, 6 (1896–9), 65–74

101 'Sul §2 del Formulario t. II: Aritmetica,' *R.d.M.*, 6 (1896–9), 75–89

102 Review of E. Schröder, *Ueber Pasigraphie* etc., *R.d.M.*, 6 (1896–9), 95–101

103 'La numerazione binaria applicata alla stenografia,' *Atti Accad. sci. Torino*, 34 (1898–9), 47–55

103.1 Reply to question no. 1132, *L'intermédiaire des mathématiciens*, 5 (1898), 23

103.2 Reply to question no. 1185, *L'intermédiaire des mathématiciens*, 5 (1898), 71–2

103.3 Question no. 1297, *L'intermédiaire des mathématiciens*, 5 (1898), 125–6

103.4 Reply to question no. 1208, *L'intermédiaire des mathématiciens*, 5 (1898), 144

1899

104 Letter to the Editor, *Period. mat.*, (2) 14 (1899), 152–3

105 'Sui numeri irrazionali,' *R.d.M.*, 6 (1896–9), 126–40

VIII, 106 *Formulaire de mathématiques*, vol. 2, section 3 (Turin: Bocca, 1899), pp. 199

106.1 Reply to question no. 1374, *L'intermédiaire des mathématiciens*, 6 (1899), 135

f *Revue de mathématiques*, vol. 6 (Turin: Bocca, 1896–9), pp. iv + 188

1900

107 'Formules de logique mathématique,' *R.d.M.*, 7 (1900–1), 1–41

108 'Additions au Formulaire,' *R.d.M.*, 7 (1900–1), 67–70

109 'Les définitions mathématiques,' *Congrès international de philosophie, Paris, 1900* (Paris, 1901), vol. 3, pp. 279–88

109' 'Definicye w matematyce,' trans. by Z. Krygowski, *Wiadom. mat.*, 6 (1902), 174–81

1901

IX, 110   *Formulaire de mathématiques*, vol. 3 (Paris: Carré & Naud, 1901), pp. viii + 231

111   *Studio delle basi sociali della Cassa nazionale mutua cooperativa per le pensioni* (Torino, 1901), pp. iv + 34

111′   *Studio delle basi sociali della Cassa mutua cooperativa per le pensioni*, 2nd ed. (Torino, 1906), pp. 34

112   (With others) 'Additions et corrections au Formulaire a.1901,' *R.d.M.*, 7 (1900–1), 85–110

113   Review of O. Stolz and I. A. Gmeiner, Theoretische Arithmetik, I, *R.d.M.*, 7 (1900–1), 112–14

114   *Relazione della Commissione per lo studio delle basi sociali della Cassa etc.* (Torino, 1901), pp. 10

115   'Memoria,' *Bollettino di notizie sul credito etc., del Ministero agr. ind. e comm.* (1901), 1200–(?)

116   *Dizionario di logica matematica*, Presented to the Congress of Secondary School Professors, Livorno, 1901, pp. 8

116′   'Dizionario di matematica, parte I, Logica matematica,' *R.d.M.*, 7 (1900–1), 160–72

116″   *Dizionario di matematica*, parte I, *Logica matematica* (Torino, 1901), pp. 16

116.1   Letter to the Editor, *Il bollettino di matematiche e di scienze fisiche e naturali. Giornale per la coltura dei maestri delle scuole elementari e degli alunni delle scuole normale*, 2 (1900–1), 254–5

117   (With others) 'Additions au Formulaire a.1901,' *R.d.M.*, 7 (1900–1), 173–84

g   *Revue de mathématiques*, vol. 7 (Turin: Bocca, 1900–1), pp. iv + 184

1902

X, 118   *Aritmetica generale e algebra elementare* (Torino: Paravia, 1902), pp. viii + 144

119   Review of C. Arzelà, *Lezioni di calcolo infinitesimale*, vol. I, part 1, *R.d.M.*, 8 (1902–6), 7–11

120   *Seconda relazione della Commissione per lo studio delle basi sociali della Cassa etc.* (Torino, 1902), pp. 28

121   Letter to the Editor (Sur les imaginaires), *Bulletin des sciences mathématiques et physiques élémentaires*, 7 (1901–2), 275–7

122   'La geometria basata sulle idee di punto e distanza,' *Atti Accad. sci. Torino*, 38 (1902–3), 6–10

123   *A proposito di alcuni errori contenuti nel disegno di legge sulle associazioni tontinarie presentato al Senato* (Torino, 1902), pp. 2

1903

124   *Sul massimo della pensione a distribuirsi dalla Cassa etc.* (Torino, 1903), pp. 20

xi, 125   *Formulaire mathématique*, vol. 4 (Turin: Bocca, 1903), pp. xvi + 407

*126-1   'De latino sine-flexione,' *R.d.M.*, 8 (1902–6), 74–83

127-126   'Principio de permanentia,' *R.d.M.*, 8 (1902–6), 84–7

*127'-1'   *De latino sine flexione – Principio de permanentia* (Torino, 1903), pp. 14

127.1   Question no. 2549, *L'intermédiaire des mathématiciens*, 10 (1903), 70

1904

*128-2   'Il latino quale lingua ausiliare internazionale,' *Atti Accad. sci. Torino*, 39 (1903–4), 273–83

129-127   'Sur les principes de la géométrie selon M. Pieri,' *Kazan Universitet, Fiziko-matematicheskoe obshchestvo, Izvīestīīa* (Bulletin de la Société physico-mathématique de Kasan), (2) 14 (1905), 92–5

*130-3   *Vocabulario de latino internationale, comparato cum Anglo, Franco, Germano, Hispano, Italo, Russo, Graeco et Sanscrito* (Torino, 1904), pp. 40

1905

131-128   *Progetto di una Cassa di riassicurazione e di una Cassa di soccorso etc.* (Torino, 1905), pp. 58

131.1   (With C. Segre) 'Relazione sulla memoria del prof. Mario Pieri: Nuovi principî di geometria projettiva complessa,' *Atti Accad. sci. Torino*, 40 (1904–5), 378–9

1906

132-129   'Sulle differenze finite,' *Atti Accad. naz. Lincei, Rend., Cl. sci. fis. mat. nat.*, (5) 15-i (1906), 71–2

133-130   'Super theorema de Cantor-Bernstein,' *Rend. Circ. mat. Palermo*, 21 (1906), 360–6

133'-130'   'Super theorema de Cantor-Bernstein et additione,' *R.d.M.*, 8 (1902–6), 136–57

*134-4   'Notitias super lingua internationale,' *R.d.M.*, 8 (1902–6), 159

h   *Revista de mathematica*, vol. 8 (Torino: Bocca, 1902–6), pp. iv + 160

*146-9   (See page numbers for titles) *Academia pro Interlingua, Discussiones*, 1 (1909–10), 1–6 (untitled); 9–13, 'Propositiones in discussione'; 57–8, 'Propositiones in discussione'; (with G. Pagliero) 77–81, ' "Discussiones" in "Progreso" '

*147-10   'Lingua de Academia,' *A.p.I. Discussiones*, 1 (1910), 91–6, 147–57, 187–91

*148-11   'Exemplo de Interlingua,' *A.p.I. Discussiones*, 1 (1910), 163–74, 191–9

    i   *Academia pro Interlingua, Discussiones*, vol. 1 (Torino, 1909–10), pp. 225

1911

149-138   'Sulla definizione di funzione,' *Atti Accad. naz. Lincei, Rend.*, *Cl. sci. fis. mat. nat.*, (5) 20-I (1911), 3–5

*150-12   'Le latin sans flexions,' *Les questions modernes*, 2 (1911), 509–12

151-139   'Le definizioni in matematica,' *Institut d'estudis catalans, Barcelona, Seccio de ciencies, Arxius*, 1 (1911–12), no. 1, 49–70

*152-13   'De passivo,' *A.p.I. Discussiones*, 2 (1911), 81–4

*153-14   *100 exemplo de Interlingua cum vocabulario Interlingua-italiano* (Torino: Bocca, 1911), pp. 16

*153'-14'   *100 exemplo de Interlingua cum vocabulario Interlingua-latino-italiano-français-english-deutsch*, 2nd ed. (Torino, 1913)

*154-15   'Una questione di grammatica razionale,' *Fourth International Congress of Philosophy, Bologna, 5–11 April 1911* (Genoa: Formiggini), vol. 2, pp. 343–8

154.1   'Observaciones etimológicas,' *Revista de la Sociedad matemática española*, 1 (1911), 143

    j   *Academia pro Interlingua, Discussiones*, vol. 2 (Torino: Bocca, 1911), pp. 132

1912

*155-16   'De derivatione,' *A.p.I. Discussiones*, 3 (1912), 20–43

156-140   'Sulla definizione di probabilità,' *Atti Accad. naz. Lincei, Rend.*, *Cl. sci. fis. mat. nat.*, (5) 21-I (1912), 429–31

157-141   'Delle proposizioni esistenziali,' *Fifth International Congress of Mathematicians, Cambridge, 22–28 August 1912* (Cambridge: University Press, 1913), vol. 2, pp. 497–500

158-142   'Contro gli esami,' *Torino nuova*, (17 August 1912), 2

159-143   'Derivata e differenziale,' *Atti Accad. sci. Torino*, 48 (1912–13), 47–69

159.1 Review of F. Castellano, *Lezioni di meccanica razionale*, seconda edizione, *Bollettino di bibliografia e storia delle scienze matematiche*, 14 (1912), 7

159.2 Preface to Alessandro Padoa, *La logique déductive dans sa dernière phase de développement* (Paris: Gauthier-Villars, 1912), pp. 106 (preface, pp. 3–4)

159.3 (With C. Somigliana) 'Relazione sulla memoria del Dr. G. Sannia: Caratteristiche multiple di un'equazione alle derivate parziali in due variabili indipendenti,' *Atti Accad. sci. Torino*, 48 (1912–13), 196

159.4 (Unsigned) 'Martin Schleyer,' *A.p.I. Discussiones*, 3 (1912), 164–70

k *Academia pro Interlingua, Discussiones*, vol. 3 (Torino: Bocca, 1912), pp. 204

1913

*160-17 'Vocabulario de Interlingua,' *A.p.I. Discussiones*, 4 (1913), 13–19

*160'-17' 'De vocabulario,' *A.p.I. Circulare*, (1924), no. 4, 9–13. Reprinted in *Schola et Vita*, 1 (1926), 191–5

161-144 'Sulla definizione di limite,' *Atti Accad. sci. Torino*, 48 (1912–13), 750–72

162-145 'Resto nelle formule di quadratura espresso con un integrale definito,' *Atti Accad. naz. Lincei, Rend., Cl. sci. fis. mat. nat.*, (5) 22-ɪ (1913), 562–9

163-146 Review of A.N. Whitehead and B. Russell, *Principia mathematica*, vols. ɪ, ɪɪ, *Bollettino di bibliografia e storia delle scienze matematiche*, 15 (1913), 47–53, 75–81

163.1 (Unsigned) 'Mario Pieri,' *A.p.I. Discussiones*, 4 (1913), 31–5

*164-18 'Quaestiones de grammatica,' *A.p.I. Discussiones*, 4 (1913), 41–4

165-147 (With E. D'Ovidio) 'Relazione sulla memoria del Dr. Vincenzo Mago: Teoria degli ordini,' *Atti Accad. sci. Torino*, 49 (1913–14), 169

l *Academia pro Interlingua, Discussiones*, vol. 4 (Torino: Bocca, 1913), pp. 64

1914

166-148 'Residuo in formulas de quadratura,' *Mathesis*, (4) 4 (1914), 5–10

*167-19 *Fundamento de Esperanto* (Cavoretto (Torino), 1914), pp. 16

167.1 'Prof. Augusto Actis,' *Revista universale* (U. Basso, ed.), 4 (1914), no. 39 (March), 22

167.2 'Prof. Louis Couturat,' *Revista universale* (U. Basso, ed.), 4 (1914), no. 40 (October), 78–9

168-149   (With C. Segre) 'Relazione intorno alla memoria del Prof. Cesare Burali-Forti, Isomerie vettoriali e moti geometrici,' *Atti Accad. sci. Torino*, 50 (1914–15), 237

1915

xiv, *169-20   *Vocabulario commune ad latino-italiano-français-english-deutsch*, 2nd ed. (Cavoretto (Torino): Academia pro Interlingua, 1915), pp. xxxii + 320

170-150   'Resto nella formula di Cavalieri-Simpson,' *Atti Accad. sci. Torino*, 50 (1914–15), 481–6

170'-150'   'Residuo in formula de quadratura Cavalieri-Simpson,' *Enseignement math.*, 18 (1916), 124–9

171-151   'Sul prodotto di grandezze,' *Bollettino di matematiche e di scienze fisiche e naturali*, 16 (1915), 99–100

172-152   'Definitione de numero irrationale secundo Euclide,' *'Mathesis' Società italiana di matematica, Bollettino*, 7 (1915), 31–5

*173-21   'Praepositiones internationale,' *World-Speech* (Marietta, Ohio), (1915), no. 37, 2–3

174-153   (With E. D'Ovidio) 'Relazione sulla memoria di G. Sannia, I limiti d'una funzione in un punto limite del suo campo,' *Atti Accad. sci. Torino*, 50 (1914–15), 968–9

175-154   'Le grandezze coesistenti di Cauchy,' *Atti Accad. sci. Torino*, 50 (1914–15), 1146–55

176-155   'Importanza dei simboli in matematica,' *Scientia*, 18 (1915), 165–73

176'-155'   'Importance des symboles en mathématique,' *Scientia*, 18 (1915), supplement, 105–14

177-156   'Le definizioni per astrazione,' *'Mathesis' società italiana di matematica, Bollettino*, 7 (1915), 106–20

178-157   'L'esecuzione tipografica delle formule matematiche,' *Atti Accad. sci. Torino*, 51 (1915–16), 279–86. Reprinted in *Bollettino di matematica*, 14 (1915–16), 121–8

*179-22   'Bello et Lingua,' *International Language; a Journal of 'Interlinguistics,'* 5 (1915), 17–19

1916

180-158   'Approssimazioni numeriche,' *Atti Accad. naz. Lincei, Rend., Cl. sci. fis. mat. nat.*, (5) 25-I (1916), 8–14

181-159   'Sul principio d'identità,' *'Mathesis' società italiana di matematica, Bollettino*, 8 (1916), 40–1 m   'Gli Stati Uniti della Terra,' *Gazzetta del popolo della sera*, Torino, 8 March 1916

1917

182-160   'Valori decimali abbreviati e arrotondati,' *Atti Accad. sci. Torino*, 52 (1916–17), 372–82. Reprinted in *Period. mat.*, (3) 14 (1917), 97–105

183-161   'Approssimazioni numeriche,' *Atti Accad. sci. Torino*, 52 (1916–17), 453–68, 513–28

1918

184-162   'Eguale,' 'Infinito,' 'Logica matematica,' 'Vettori,' *Dizionario cognizioni utili* (and supplement) (Torino: Unione Tipografico-Editrice Torinese, 1917–19)

184'-162'   'Eguale,' *Bollettino di matematica*, 15 (1917–18), 195–8

184''-162''   'Vettori,' *Bollettino di matematica*, 16 (1918–19), 157–8

185-163   *Tavole numeriche* (Unione Tipografico-Editrice Torinese, 1918), pp. 32 (11th reprinting, 1963)

186-164   'Interpolazione nelle tavole numeriche,' *Atti Accad. sci. Torino*, 53 (1917–18), 693–716

187-165   'Resto nelle formule di interpolazione,' *Scritti offerti ad E. D'Ovidio* (Torino: Bocca, 1918), pp. 333–5

188-166   (With T. Boggio) 'Matteo Bottasso,' '*Mathesis' società italiana di matematica, Bollettino*, 10 (1918), 87–8

1919

189-167   'Sulla forma dei segni di algebra,' *Giorn. mat. finanz.*, 1 (1919), 44–9

190-168   'Filiberto Castellano,' '*Mathesis' società italiana di matematica, Bollettino*, 11 (1919), 62–3

191-169   'Risoluzione graduale delle equazioni numeriche,' *Atti Accad. sci. Torino*, 54 (1918–19), 795–807

192-170   'Conferenze matematiche,' *Annuario dell'Università di Torino* (1919–20), 245–6

1921

193-171   'Le definizioni in matematica,' *Period. mat.*, (4) 1 (1921), 175–89

194-172   'Area de rectangulo,' *Rassegna di matematica e fisica* (Rome, Istituto fisico-matematico G. Ferraris), 1 (1921), 200–3

195-173   Review of T. Boggio, *Calcolo differenziale, con applicazioni geometriche, Esercitazioni matematiche* (Circolo matematico, Catania), 1 (1921), 198–201

1922

196-174   'Calculo super calendario,' *A.p.I. Circulare*, (1 January 1922), 2–3, 6–7

196'-174'   'Calculo super calendario,' *Urania; notizie di astronomia, meteorologia, geologia, mineralogia, chimica e fisica*, 11 (1922), 20–4

196''-174''   'Calculo super calendario,' *Schola et vita*, 3 (1928), 4–6

197-173   'Operazioni sulle grandezze,' *Atti Accad. sci. Torino*, 57 (1921–2), 311–31

197'-175'   'Operationes super magnitudines,' *Rassegna di matematica e fisica* (Roma, Istituto fisico-matematico G. Ferraris), 2 (1922), 269–83

*198-23   Review of A.L. Guérard, A Short History of the International Language Movement, *A.p.I. Circulare*, (1922), no. 2, 5–11

*199-24   'Regulas pro Interlingua,' *A.p.I. Circulare*, (1922), no. 3, 1–3

*200-25   'Lingua internationale ante Societate de Nationes,' *A.p.I. Circulare*, (1922), no. 4, 1–4

201-176   'Theoria simplice de logarithmos,' *Wiadom. mat.*, 26 (1922), 53–5

201.1   Letter to the Editor, *La matematica elementare*, 1 (1922), 76

201.2   'Problemi pratici,' *La matematica elementare*, 1 (1922), 88–9

1923

202-177   Review of A. Nantucci, *Il concetto di numero, e le sue estensioni, Archeion; archivio di storia della scienza*, 4 (1923), 382–3

202.1   Review of Feder and Schneeberger, *Vollständiges Wurzelwörterbuch Ido-Deutsch* (Lüsslingen, 1919) and S. Auerbach, *Wörterbuch Deutsch-Ido* (Leipzig, 1922), *A.p.I. Circulare*, (1923), no. 1, 6–8

202.2   Review of Mondo, *Revuo por linguo Internaciona Ido, A.p.I. Circulare*, (1923), no. 3 (September), 7–8

202.3   'Discussione,' *A.p.I. Circulare*, (1923), no. 4 (December), 2–3

1924

xv, 203-178   *Giochi di aritmetica e problemi interessanti* (Torino: Paravia, 1924), pp. 64 (Several later editions)

204-179   'I libri di testo per l'aritmetica nelle scuole elementari,' *Period. mat.*, (4) 4 (1924), 237–42

*205-26   *Interlingua (Historia – Regulas pro Interlingua – De vocabulario – Orthographia – Lingua sine grammatica)* (Cavoretto (Torino), 1924), pp. 24 (2nd ed., 1925; 3rd ed., 1927)

206-180 'De aequalitate,' *Proceedings of the International Congress of Mathematicians, Toronto, 1924* (Toronto: University of Toronto Press, 1928), vol. 2, 988–9
206'-180' 'De aequalitate,' *A.p.I. Circulare*, (1924), no. 5, 8–11
XVI, *207-27 *Academia pro Interlingua, Circulares* (Torino, 1909–24)

1925

207.1 Letter to the Editor (A proposito dell'articolo di L. Siriati), *Bollettino di matematica*, (2) 3 (1925), 129
207.2 'International Auxiliary Language Association,' *A.p.I. Circulare*, (1925), no. 5 (August), 76–7
*208-28 'Pro historia de Interlingua, §1. Volapük,' *A.p.I.*, (1925), no. 5, 81–5
n *Academia pro Interlingua* (Cavoretto (Torino), 1925), pp. 112

1926

*209-29 'Pro historia de Interlingua, §2. Academia in periodo 1893–1908,' *A.p.I.*, (1926), no. 2, 33–4
*210-30 'Pro historia de Interlingua, §3. Latino simplificato,' *A.p.I.*, (1926), no. 4, 73–6
*211-31 'Sinense,' *A.p.I.*, (1926), no. 6, 129–30
212-181 'Quadrato magico,' *Schola et vita*, 1 (1926), 84–7
213-182 'Jocos de arithmetica,' *Schola et vita*, 1 (1926), 166–73
o *Academia pro Interlingua* (Cavoretto (Torino), 1926), pp. 144

1927

*214-32 Review of E.S. Pankhurst, *Delphos the Future of International Language, A.p.I.*, (1927), no. 1, 14–20
215-183 'De vocabulo matematica,' *Rivista di matematica pura ed applicata per gli studenti delle scuole medie*, 2 (1927), 212
216-184 'Sulla riforma del calendario,' *Atti Accad. sci. Torino*, 62 (1926–7), 566–8
217-185 'Historia et reforma de calendario,' *A.p.I.*, (1927), no. 3, 49–55
218-186 'Vocabulario matematico,' *Rivista di matematica pura ed applicata per gli studenti delle scuole medie*, 3 (1927), 270–2
p Academia pro Interlingua (Cavoretto (Torino), 1927), pp. 112

1928

219-187 'Gio. Francesco Peverone ed altri matematici piemontesi ai tempi di Emanuele Filiberto,' *Studi pubblicati dalla R. Università di Torino nel* IV *centenario della nascita di Emanuele Filiberto* (Torino: R. Università di Torino), pp. 183–9

220-188   'Historia de numeros,' *Schola et vita*, 3 (1928), 139–42. Reprinted in *Archeion; archivio di storia della scienza*, 9 (1928), 364–6

220.1   'Ad omne interlinguista collega et amico, gratias,' *Schola et vita*, 3 (1928), 201

221-189   'Interessante libro super calculo numerico,' *Schola et vita*, 3 (1928), 217–19

222-190   (With Cesarina Boccalatte) 'La geometria basata sulle idee di punto e angolo retto,' *Atti Accad. sci. Torino, Cl. sci. fis. mat. nat.*, 64-ɪ (1928–29), 47–55

1929

*223-33   'Vocabulos internationale,' *Schola et vita*, 4 (1929), 58–61

*224-34   'Volapük post 50 anno,' *Schola et vita*, 4 (1929), 225–33

224.1   Letter to the Editor, *Time and Tide* (London, 5 July 1929). Reprinted in *Schola et vita*, 4 (1929), 247–8

225-191   'Monete italiane nel 1929,' *Giornale di matematica e fisica*, 3 (1929), no. 4

1930

*226-35   'Studio de linguas,' *Rendiconti della Unione professori*, Milano, (April 1930), 1–2. Reprinted in *Schola et vita*, 5 (1930), 81–4

*227-36   'Quaestiones de interlingua: ablativo aut nominativo,' *Schola et vita*, 5 (1930), 138–40

228-192   (With Fausta Audisio) 'Calcolo di π colla serie di Leibniz,' *Atti Accad. naz. Lincei, Rend., Cl. sci. fis. mat. nat.*, (6) 11-ɪ (1930), 1077–80

*229-37   'Algebra de grammatica,' *Schola et vita*, 5 (1930), 323–36

1931

230-193   'Jocos de arithmetica,' *Rendiconti della Unione professori*, Milano, (October 1931), 50–1

*231-38   'Libertate et Unione,' *Schola et vita*, 6 (1931), 323–5

Bibliography on the
life and works of Giuseppe Peano

BOOKS

Ugo Cassina edited the three-volume edition of Peano's *Opere scelte* (Rome: Edizioni Cremonese, 1957–59) and contributed a very valuable introduction, biographical sketch, and set of introductory notes to the works selected. He also wrote an introduction and notes for the facsimile reprint of the *Formulario mathematico* (Rome: Edizioni Cremonese, 1960). Two issues of *Schola et vita*, edited by Cassina, were devoted to Peano: the supplement of 27 August 1928, in honour of Peano's seventieth birthday; and no. 3 of volume 7 (1932, pp. 97–160), a commemorative issue following the death of Peano. The volumes *Critica dei principî della matematica e questioni di logica* (Rome: Edizioni Cremonese, 1961) and *Dalla geometria egiziana alla matematica moderna* (Rome: Edizioni Cremonese, 1961) contain reprints of articles by Cassina. Those treating of Peano are listed below.

   In Memoria di Giuseppe Peano (Cuneo, 1955) contains articles by eight authors, which are included in the list below.

ARTICLES

Ascoli, Guido, (untitled necrology of Peano), *La ricerca scientifica*, 3 (1932),
      592–3
— 'I motivi fondamentale dell'opera di Giuseppe Peano,' in *In Memoria*,
      pp. 23–30
Barone, Francesco, 'Un'apertura filosofica della logica simbolica peaniana,' in
      In Memoria, pp. 41–50
Boggio, Tommaso, 'Giuseppe Peano,' *Atti Accad. sci. Torino, Cl. sci. fis. mat.
      nat.*, 68 (1932–3), 436–46
— 'Il calcolo geometrico di Peano,' in *In Memoria*, pp. 65–9
Botto, Costantino, 'Un'autentica gloria cuneese e italiano: il matematico
      Giuseppe Peano, Cuneo 1858 – Torino 1932,' *Annuario del Istituto tecnico
      di Cuneo*, 1933–4, pp. 5–24

Carruccio, Ettore, 'Spunti di storia delle matematiche e della logica nell'opera di G. Peano,' in *In Memoria*, pp. 103–14

Cassina, Ugo, 'In occasione de septuagesimo anno de Giuseppe Peano,' supplement to *Schola et vita*, 27 August 1928, pp. 5–28

— 'Vita et opera de Giuseppe Peano,' *Schola et vita*, 7 (1932), 117–48

— 'Su la logica matematica di G. Peano,' *Boll. Un. mat. ital.*, 12 (1933), 57–65 (Also in *Critica*)

— 'L'œuvre philosophique de G. Peano,' *Revue de métaphysique et de morale*, 40 (1933), 481–91

— 'L'opera scientifica di Giuseppe Peano,' *Rendiconti del Seminario matematico e fisico di Milano*, 7 (1933), 323–89 (Also in *Dalla geometria*)

— 'Parallelo fra la logica teoretica di Hilbert e quella di Peano,' *Period. mat.*, (4) 17 (1937), 129–38 (Also in *Critica*)

— 'L'area di una superficie curva nel carteggio inedito di Genocchi con Schwarz ed Hermite,' *Istituto Lombardo, Accademia di scienze e lettere, Rendiconti, Scienze*, 83 (1950), 311–28 (Also in *Dalla geometria*)

— 'Il concetto di linea piana e la curva di Peano,' *Rivista di matematica della Università di Parma*, 1 (1950), 275–92

— 'Alcune lettere e documenti inediti sul trattato di Calcolo di Genocchi-Peano,' *Istituto Lombardo, Accademia di scienze e lettere, Rendiconti, Scienze*, 85 (1952), 337–62 (Also in *Dalla geometria*)

— 'L'idéographie de Peano du point de vue de la théorie du langage,' *Rivista di matematica della Università di Parma*, 4 (1953), 195–205 (Italian translation in *Critica*)

— 'Sulla critica di Grandjot all'aritmetica di Peano,' *Boll. Un. mat. ital.*, (3) 8 (1953), 442–7 (Also in *Critica*)

— 'Su l'opera filosofica e didattica di Giuseppe Peano' (in *Critica*. This was a talk given in Cuneo, 6 December 1953)

— 'Sul *Formulario mathematico* di Peano,' in *In Memoria* (Also in *Critica*)

— 'Storia ed analisi del "Formulario completo" di Peano,' *Boll. Un. mat. ital.*, (3) 10 (1955), 244–65, 544–74 (Also in *Dalla geometria*)

— 'Un chiarimento sulla biografia di G. Peano,' *Boll. Un. mat. ital.*, (3) 12 (1957), 310–12

— 'Su un teorema di Peano e il moto del polo,' *Istituto Lombardo, Accademia di scienze e lettere, Rendiconti, Scienze*, 92 (1958), 631–55

Couturat, Louis, 'La logique mathématique de M. Peano,' *Revue de métaphysique et de morale*, 7 (1899), 616–46

Dellacasa, Luciano, 'Infantia et juventute de Giuseppe Peano,' *Schola et vita*, 8 (1933), 141–4

Dickstein, Samuel, 'Peano as Historian of Mathematics' (in Polish), *Wiadom. mat.*, 36 (1934), 65–70

Feys, Robert, 'Peano et Burali-Forti précurseurs de la logique combinatoire,' *Actes du* xième *Congrès international de philosophie*, vol. 5 (Amsterdam, Louvain, 1953), pp. 70–2

Frege, Gottlob, 'Ueber die Beggriffsschrift des Herrn Peano und meine eigene,' *Berichte über die Verhandlungen der Königlich Sächsischen Gesellschaft der Wissenschaften zu Leipzig, mathematisch-physikalische Klasse*, 48 (1896), 361–78

Gabba, Alberto, 'La definizione di area di una superficie curva ed un carteggio inedito di Casorati con Schwarz e Peano,' *Istituto Lombardo, Accademia di scienze e lettere, Rendiconti, Scienze*, 91 (1957), 857–83

Geymonat, Ludovico, 'I fondamenti dell'aritmetica secondo Peano e le obiezioni "filosofiche" di B. Russell,' in *In Memoria*

— 'Peano e le sorti della logica in Italia,' *Boll. Un. mat. ital.*, (3) 14 (1959), 109–18

Gliozzi, Mario, 'Giuseppe Peano,' *Archeion; archivio di storia della scienza*, 14 (1932), 254–5

Jourdain, P.E.B., 'The Development of the Theories of Mathematical Logic and the Principles of Mathematics,' *Quarterly Journal of Pure and Applied Mathematics*, 43 (1912), 219–314 (pp. 270–314 treat of Peano's work)

Kennedy, H.C., 'An Appreciation of Giuseppe Peano,' *Pi Mu Epsilon Journal*, 3 (1960), 107–13

— 'The Mathematical Philosophy of Giuseppe Peano,' *Philosophy of Science*, 30 (1963), 262–6

— 'Giuseppe Peano at the University of Turin,' *The Mathematics Teacher*, 61 (1968), 703–6

Kozlowski, W.M., 'Commemoration of Giuseppe Peano' (in Polish), *Wiadom. mat.*, 36 (1934), 57–64

Levi, Beppo, 'L'opera matematica di Giuseppe Peano,' *Boll. Un. mat. ital.*, 11 (1932), 253–62

— 'Intorno alle vedute di G. Peano circa la logica matematica,' *Boll. Un. mat. ital.*, 12 (1933), 65–8

— L'opera matematica di Giuseppe Peano,' in *In Memoria*

Medvedev, F.A., 'G. Peano's Functions of a Set' (in Russian), *Istoriko-Matematičeskie Issledovaniya*, 16 (1965), 311–23

Natucci, A., 'In memoria di G. Peano,' *Bollettino di matematica*, (2) 11 (1932), 52–6

Nidditch, Peter, 'Peano and the Recognition of Frege,' *Mind*, 72 (1963), 103–10

Padoa, Alessandro, 'Il contributo di G. Peano all'ideografia logica,' *Period. mat.*, (4) 13 (1933), 15–22

— 'Ce que la logique doit à Peano,' *Actualités scientifiques et industrielles*, 395 (Paris, 1936), 31–7

Segre, Beniamino, 'Peano ed il Bourbakismo,' in *In Memoria*

Stamm, Edward, 'Jozef Peano' (in Polish), *Wiadom. mat.*, 36 (1934), 1–56

Tripodi, Angelo, 'Considerazioni sull'epistolario Frege-Peano,' *Boll. Un. mat. ital.*, (4) 3 (1970), 690–8

Vacca, Giovanni, 'Lo studio dei classici negli scritti matematici di Giuseppe Peano,' *Atti della Società italiana per il progresso delle scienze*, 21 (1932), II, 97–9

Vailati, Giovanni, 'La logique mathématique et sa nouvelle phase de développement dans les écrits de M. J. Peano,' *Revue de métaphysique et de morale*, 7 (1899), 86–102

Vivanti, Giulio, 'Giuseppe Peano,' *Istituto Lombardo di scienze e lettere, Rendiconti*, (2) 65 (1932), 497–8

TRANSLATIONS OF SELECTED PUBLICATIONS

I

On the integrability of functions (1883)*

We begin with Peano's first publication in analysis. It contains, implicitly, his notion of integral, given explicitly in the Calcolo differenziale of 1884. Near the end he introduces the concept of area, using the notions of least upper bound and greatest lower bound. This concept is analogous to those of length and volume which are defined explicitly in the Applicazioni geometriche of 1887.

The existence of the integral of functions of one variable is not always proved with the rigour and simplicity desired in such questions. Geometrical considerations are brought in and, in general, the mode of reasoning of the major textbook writers is not satisfactory. The analytic proofs are long and complicated; conditions are introduced in them which are either too restrictive or in part useless. I propose, in the present note, to prove the existence of the integral by introducing a quite simple condition of integrability. The reasoning will be analytic, but every part can be given a geometric interpretation.

1

Let $y = f(x)$ be a function of x defined on an interval (a, b). Suppose a and b are finite and that y is bounded on the interval. Let A and B be the least upper bound (l.u.b.) and the greatest lower bound (g.l.b.), respectively, of y on the interval. Partition the interval (a, b) into parts of length $h_1, h_2, ..., h_n$. Let y_s be any value whatever assumed by y as x varies in the interval h_s, and form the sum

$$u = h_1 y_1 + h_2 y_2 + ... + h_n y_n = \sum h_s y_s.$$

If, with the indefinite diminishing of all the h, u tends toward a limit S,[1]

---

* 'Sulla integrabilità delle funzioni,' *Atti Accad. sci. Torino*, 18 (1882–3), 439–46 [5].

1 By 'u tends toward a limit S' we mean that, having fixed a quantity ε as small as we like, we may determine a σ such that for every partition of (a, b) for which the length h of each interval is less than σ, and for every choice of the y_s in the intervals, we always have $|S - u| < \varepsilon$.

the function is said to be integrable on the interval (a, b) and this limit is said to be the value of the integral

$$\int_a^b f(x)dx.$$

2

Let p_s and q_s be the l.u.b. and the g.l.b., respectively, of y on the interval h_s. Set

$$P = \sum h_s p_s, \qquad Q = \sum h_s q_s.$$

Since $A \geqslant p_s \geqslant y_s \geqslant q_s \geqslant B$, multiplying by h_s and summing, we have

$$A(b - a) \geqslant P \geqslant u \geqslant Q \geqslant B(b - a).$$

Here, P and Q are upper and lower bounds, respectively, for u.

By varying the partition of (a, b), we may vary P and Q. But every value of P is greater than or equal to every value of Q. In fact, let $h_1, h_2, ..., h_n$ and $h_1', h_2', ..., h_n'$ be the lengths of the respective subintervals of two partitions of (a, b), and consider the partition formed by superimposing these two partitions. Label the lengths of the subintervals of the new partition $k_1, k_2, ..., k_m$. Every interval k_α is contained in a h_β and a h_γ', and every h and every h' is equal to the sum of one or more intervals k. Since from the original partitions we had

$$P = \sum_{r=1}^{n} h_r p_r, \qquad Q' = \sum_{s=1}^{n'} h_s' q_s',$$

we now have, after appropriate substitutions,

$$P = \sum_{\alpha=1}^{m} k_\alpha p_\beta, \qquad Q' = \sum_{\alpha=1}^{m} k_\alpha q_\gamma', \qquad \text{and } P - Q' = \sum_\alpha k_\alpha (p_\beta - q_\gamma').$$

But p_β is the l.u.b. of y on the interval h_β, which contains k_α, and q_γ' is the g.l.b. of y on the interval h_γ', which contains k_α. Hence

$$p_\beta > q_\gamma', \qquad P - Q' > 0, \qquad \text{and } P > Q', \qquad \text{Q.E.D.}$$

Thus the quantities P, which are all finite, have a greatest lower bound M and the quantities Q have a least upper bound N, such that

$$P \geqslant M \geqslant N \geqslant Q.$$

3

If $f(x)$ is integrable, given an arbitrarily small ε, we can find a quantity σ such that for every partition of (a, b), for which every $h < \sigma$, u is always between $S + \varepsilon$ and $S - \varepsilon$. The values of P and Q corresponding to these partitions are also between $S + \varepsilon$ and $S - \varepsilon$, because u can assume values as close as we like to every value of P and Q. The quantities M and N, being between P and Q, will also be between $S + \varepsilon$ and $S - \varepsilon$. Thus, we will have $M = N = S$, because M, N, S are constant quantities and ε is as small as we like.

Hence, if the function $f(x)$ is integrable:

1. The quantities M and N are equal, and their common value is equal to the value of the integral.

2. The quantities P and Q tend toward S as the intervals diminish.

3. The difference between two values which P and Q may assume corresponding to the same or to different partitions of (a, b) may be made as small as we like by taking the intervals of the two partitions sufficiently small.

If in this last proposition we suppose that P and Q correspond to the same partition, by setting $p_s - q_s = d_s$ (the oscillation of y in the interval h_s) and $D = \sum h_s d_s$, we will have

$$P - Q = \sum h_s(p_s - q_s) = D.$$

Hence, if $f(x)$ is integrable, D has zero as its limit as the intervals h diminish indefinitely.

The preceding conditions, necessary for integrability, are not mutually independent, as the following quite simple theorem shows.

4

THEOREM The function $f(x)$ is integrable on the interval (a, b) if $M = N$, and the common value S is the value of the integral.

Let $h_1, h_2, ..., h_n$ be any partition whatever of (a, b) such that every h is less than σ, a quantity to be determined. Let $u = \sum h_s y_s$.

Since the g.l.b. of the values of P is S, given an arbitrary ε we can make a partition $h_1', h_2', ..., h_n'$ of (a, b) for which, if we let $P' = \sum h_s' p_s'$, we have $P' - S < \varepsilon$. Now consider the partition of (a, b) derived from the superposition of the preceding partitions, and an interval k_α of this contained in h_β and in h_γ'. We will have $P' = \sum_\alpha k_\alpha p_\gamma'$, $u = \sum_\alpha k_\alpha y_\beta$, and $P' - u = \sum k_\alpha(p_\gamma' - y_\beta)$. Now, several of the intervals h_β can each be contained in some interval h_γ'. For them, $p_\gamma' > y_\beta$, and the corresponding terms in $P' - u$ are positive. The other intervals h_β contain some point of the second division. There are less than n' of them and, since $h_\beta < \sigma$, their total length is less than $n'\sigma$. To these intervals can correspond, in $P' - u$, negative terms, but, seeing that $p_\gamma' - y_\beta < A - B$, their sum will be less numerically than $n'\sigma(A - B)$. From this we have

$$P' - u > -n'\sigma(A - B), \qquad \text{or } S + \varepsilon + n'\sigma(A - B) > u.$$

Analogously, since S is the least upper bound of the values of Q, we can find a partition $h_1'', h_2'', ..., h_{n'}''$ for which, setting $Q'' = \sum h'' q''$, we have $S - Q'' < \varepsilon$. Thus, considering the quantity $u - Q''$, we may show in the same way that

$$u > S - \varepsilon - n''\sigma(A - B).$$

Now, given an arbitrarily small α, we may in the preceding argument suppose that $\varepsilon < \alpha/2$, and that $n'\sigma(A - B)$ and $n''\sigma(A - B)$ are less than $\alpha/2$, because it is sufficient to take $\sigma < \alpha/[2n'(A - B)]$ and $\sigma < \alpha/[2n''(A - B)]$. Then

$$S + \alpha > u > S - \alpha,$$

or, fixing a quantity α as small as we like, we may determine σ such that for every partition of (a, b) for which each $h < \sigma$, and for any set of values of the y_s, we always have $|S - u| < \alpha$, and hence u tends toward the limit S as the h decrease indefinitely, Q.E.D.

5

From the preceding theorem we may deduce another condition of integrability:

THEOREM The function $f(x)$ is integrable if, given an arbitrarily small ε, we can find a value of P and a value of Q (corresponding, or not, to the same partition of (a, b)) whose difference is less than ε. Between these two values is the value of the integral.

Indeed, M and N being between P and Q, and their difference being less than ε, we will have $M = N$, as in the hypothesis of the preceding theorem. The inverse proposition is also true, as was seen in §3.

If, in the statement of this last theorem, we make the unnecessary hypothesis that P and Q correspond to the same partition of (a, b), then recalling that $P - Q = D$, we have:

THEOREM The function $f(x)$ is integrable if the greatest lower bound of the absolute values of D is zero.[2]

2 The simple criterion of integrability stated in this theorem has already been proved in *Fondamenti per la teoria delle funzioni di variabili reali* [Pisa, 1878, p. 237] of U. Dini, but there it is deduced as a consequence of a rather long argument, which cannot be held to be elementary. In fact, the illustrious author deduces it from the criterion '$f(x)$ is integrable if lim $D = 0$ as all the h decrease,' in which unnecessary concepts (such as that of limit) are included; and in his *Lezioni di analisi infinitesimale* (Pisa, 1877–8), Dini confines himself to proving this criterion. Pasch also, in his *Einleitung in die Differential- und Integralrechnung* (Leipzig, 1882, p. 95), limits himself to this criterion.

From the preceding criteria we may easily deduce that stated by Riemann (*Ges. math. Werke* [Leipzig, 1876], p. 226) [Dover reprint, 1953, of the second edition (1892), p. 239, 'Ueber den Begriff eines bestimmten Integrals und den Umfang seiner Gültigkeit'].

In order to complete my treatment of the preceding material, I give the following theorem also:

THEOREM Every continuous function is integrable.

For, let an arbitrarily small ε be fixed. We may determine a quantity h_1 such that for every value of x between a and $a + h_1 = a_1$ we have $|f(a) - f(x)| < \varepsilon$; then a quantity h_2 such that for every value of x between a_1 and $a_1 + h_2 = a_2$ we have $|f(a_1) - f(x)| < \varepsilon$, and so on. We have, in this way, a sequence of increasing quantities $a, a_1, a_2, \ldots$ I assert that they can increase in a way to reach b. In fact, suppose they tend toward a limit $c \leqslant b$. Since $f(x)$ is continuous also for $x = c$, I can determine an interval $(c - \alpha, c)$ such that for every value of x in it we have $|f(x) - f(c)| < \varepsilon/2$. Since c is the least upper bound of the $a, a_1, a_2, \ldots$, there will exist a quantity of the sequence a_i, where i is finite, in the interval $(c - \alpha, c)$. From this we have $|f(a_i) - f(c)| < \varepsilon/2$, and supposing x is in the interval (a_i, c) and hence also in the interval $(c - \alpha, c)$, we have $|f(x) - f(c)| < \varepsilon/2$ and $|f(a_i) - f(x)| < \varepsilon$. Thus we may assume that $a_{i+1} = c$. That is to say, the value c can be reached, and then surpassed by the sequence $a, a_1, a_2, \ldots$, or this sequence of quantities can effectively reach b. Let $a, a_1, a_2, \ldots, a_{n-1}, b$ be a sequence of quantities such that in every interval $h_s = a_s - a_{s-1}$ we have $|f(a_{s-1}) - f(x)| < \varepsilon$. We then have $p_s - q_s < 2\varepsilon$, $D < 2\varepsilon(b - a)$, and hence D can be made as small as we like, because ε can be taken arbitrarily small, and thus the function is integrable.[3]

THEOREM Every bounded function which is discontinuous for a finite number of values of x in the interval (a, b) is integrable.

Indeed, if $f(x)$ becomes discontinuous for $x = c$ between a and b, we partition the interval $(a, c - \varepsilon)$, on which the function is continuous, into parts in such a way that the corresponding value of D is less than α, and the interval $(c + \varepsilon', b)$ into parts in such a way that the corresponding value of D is less than β. The result is that (a, b) will be decomposed into parts, and the value of D corresponding to this partition will be such that

$$D < \alpha + \beta + (\varepsilon + \varepsilon')(A - B)$$

---

3 Professor Lipschitz, in *Differential- und Integralrechnung* (Bonn, 1880, p. 91), in order to prove the existence of the integral, supposes from the beginning the continuity of $f(x)$, and besides, on page 97, makes a *new* hypothesis, equivalent to uniform continuity, which is clearly unnecessary. He says in fact: 'For the present we add the further hypothesis that for every pair of values x and $x + h$ lying between a and b, the numerical value of the difference $f(x + h) - f(x)$ remains smaller than an arbitrary quantity λ as soon as the numerical value of h diminishes to less than a certain quantity δ.' [*Differential- und Integralrechnung* is volume II of Rudolf Lipschitz's *Lehrbuch der Analysis*. I have translated this passage from the original. Peano quoted it incorrectly in his footnote.]

and since ε, ε', α, and β may be taken as small as we like, the greatest lower bound of the values of D is zero, and the function is integrable.

Many authors demonstrate the existence of the integral using geometrical considerations, but these arguments are not entirely satisfactory. Moreover, they exclude some integrable functions from consideration, and usually consider the area of the figure without defining it. It seems to me that the area, considered as a quantity, of a curvilinear plane figure is precisely one of those geometrical magnitudes which, like the length of the arc of a curve, is often conceived, or thought to be conceived, by our mind quite clearly, but which, before being introduced into analysis, needs to be well defined.[4] This seems to me most important in the case of area, for it is on this concept that elementary texts usually base other proofs.

Now, given a figure with a simple form, the most natural way to conceive its area is to imagine two sets of polygons such that each polygon of the first set contains the given figure in its interior and each polygon of the second set is contained in the interior of the given figure. The areas of the first have a greatest lower bound; the areas of the second have a least upper bound. If these bounds coincide, their common value is the area of the given figure, a well-determined quantity, which can be calculated to as close an approximation as one may wish. If, however, these two bounds happen not to be equal, then the concept of area should be excluded in this case. Hence, before speaking of the area of a figure, it is necessary to first verify the equality of the two bounds, which is none other than the condition of integrability given above.

It is just this equality which Professor Serret proves in his *Cours de calcul différentiel et intégral* (Paris, 1879), at no. 10, where the area is bounded by the curve $y = f(x)$, by two ordinates, and the x-axis. But in this proof one must first of all make the hypotheses that $f(x)$ is continuous (page 12, line 16) and that it does not have an infinite number of oscillations in the interval considered (page 13, line 1). Serret does not make these hypotheses and only later defines continuity of functions. Apart from all this, the proof is still not exact. Indeed, recourse is had in it to principle no. 9, which was stated and proved in vague and indefinite terms; nor does it seem easy, with the few definitions and propositions given by the author, to make the reasoning of nos. 9 and 10 rigorous. In any event,

4 'I count among these imperfections [of geometry] the obscurity of the first notions of geometrical magnitudes, of the way the measurement of these magnitudes is conceived ...' N.I.Lobachevsky, *Geometrische Untersuchungen zur Theorie der Parellellinien*, 1840. [This sentence occurs in the first paragraph. Peano quoted the French translation of G.J.Hoüel. I have translated the original German as reprinted in the *Collection complètes des œuvres géométriques*, vol. II (1886), p. 553.]

it is certain that he considers as infinitesimals quantities of the form $f(x + h) - f(x)$, where both x and h are variable. In other words, he allows in the proof that, having arbitrarily fixed ε, a quantity σ may be determined such that for every value of $h < \sigma$, and for every value of x in the interval considered, we always have $|f(x + h) - f(x)| < \varepsilon$. It is true that if $f(x)$ is continuous, then it satisfies the preceding condition (of uniform continuity), but this is a theorem which needs to be proved.

II
Differential calculus and
fundamentals of integral calculus (1884)*

Seven excerpts from the Calcolo differenziale *of 1884 are included*:

1. *An analytic expression for the function which for rational x equals* 0, *and for irrational x equals* 1 (*p. xii*).

2. *A counterexample to a theorem of Abel on convergent series* (*p. xvii*).

3. *A generalization of the mean value theorem* (*p. xxii*).

4. *A counterexample to a criterion of Serret for maxima and minima* (*pp. xxix–xxx*).

5. *An example of a function whose partial derivatives do not commute* (*p. 174*).

6. *Maxima and minima of functions of several variables* (*pp. 195–9*).

7. *A new definition of the definite integral* (*p. 298*).

P.G.L. Dirichlet had already considered the function discussed in excerpt 1 without, however, giving an analytic expression for it.

1 AN ANALYTIC EXPRESSION FOR THE FUNCTION WHICH FOR RATIONAL x EQUALS 0, AND FOR IRRATIONAL x EQUALS 1

We give the following as a remarkable example of discontinuity. Let

$$\varphi(x) = \lim_{t=0} \frac{x^2}{x^2 + t^2}.$$

We then have $\varphi(x) = 1$ if $x \gtrless 0$, and $\varphi(0) = 0$. Therefore the function of x

$$\lim \varphi(\sin n!\pi x),$$

the limit being obtained by giving n indefinitely increasing positive integral values, has the value 0 if x is rational, and the value 1 if x is irrational.

2 A COUNTEREXAMPLE TO A THEOREM OF ABEL ON CONVERGENT SERIES

The affirmation of Abel, that it is 'quite true' that 'the series cannot be convergent if the product nu_n is not zero for $n = \infty$,' must be interpreted

* Excerpts from Angelo Genocchi, *Calcolo differenziale e principii di calcolo integrale, pubblicato con aggiunte dal Dr. Giuseppe Peano* (Turin: Bocca, 1884) [8].

in the sense that the series cannot be convergent if nu_n tends toward a non-zero limit. It may happen, however, that the series is convergent even though nu_n tends toward no limit. Thus, for example, consider the series in which the terms of index

$$1, 2^3, 3^3, ..., m^3, ...$$

are respectively

$$1, \frac{1}{2^2}, \frac{1}{3^2}, ..., \frac{1}{m^2}, ...$$

and the other terms are such as to form by themselves any arbitrary convergent series. The product nu_n, if n is a cube $= m^3$, will have the value $m = \sqrt[3]{n}$; and making n increase indefinitely makes nu_n assume values as large as we like.

3 A GENERALIZATION OF THE MEAN VALUE THEOREM

If the $n + 1$ functions $f_0(x), f_1(x), ..., f_n(x)$ have derivatives up to order $n - 1$ for the values of x in an interval we wish to consider, we have

$$\begin{vmatrix} f_0{}^{(n-1)}(u) & f_1{}^{(n-1)}(u) & ... & f_n{}^{(n-1)}(u) \\ f_0(x_1) & f_1(x_1) & ... & f_n(x_1) \\ f_0(x_2) & f_1(x_2) & ... & f_n(x_2) \\ . \quad . \quad . \quad . \quad . \quad . \quad . \quad . \quad . \quad . \quad . \quad . \\ f_0(x_n) & f_1(x_n) & ... & f_n(x_n) \end{vmatrix} = 0,$$

where u is a mean value among $x_1, x_2, ..., x_n$. In this formula, if we let $f_0(x) = f(x)$ and let the successive functions be the successive powers $0, 1, 2, ...$ of x, we get the formula of no. 87 [this formula generalizes the preceding formula and that of Taylor].

More generally, the determinant whose rows are taken from $f(a)$, $f'(a), ..., f^{(\alpha)}(a), f(b), f'(b), ..., f^{(\beta)}(b), f(l), f'(l), ..., f^{(\lambda)}(l), f^{(n)}(u)$, by giving to the letter f various indices, where $n = (\alpha + 1) + (\beta + 1) + ... + (\lambda + 1) - 1$ and u is an appropriate value of x (intermediate among $a, b, ..., l$), has the value zero.

4 A COUNTEREXAMPLE TO A CRITERION OF SERRET FOR MAXIMA AND MINIMA

The criterion given by Serret (*Cours de calcul différentiel et intégral* [Paris, 1879], p. 219), 'there is a maximum or a minimum if, for the values of $h, k, ...$ which make d^2f and d^3f zero, d^4f constantly has the sign $-$ or the sign $+$,' is not exact.

In order to see the inexactness of this proposition, consider the entire function

$$f(x, y) = (y^2 - 2px)(y^2 - 2qx),$$

where $p > q > 0$. Letting $x_0 = 0$, $y_0 = 0$, we have

$$f(h, k) = 4pqh^2 - 2(p + q)hk^2 + k^4.$$

The second-degree term is positive for all values of h and k, except for the value $h = 0$, for which the third-degree term becomes zero; and the fourth-degree term is positive. Hence, according to the criterion of Serret, $f(x, y)$ is minimal for $x = 0$. But we can easily assure ourselves that this is not so. To this end, let $y^2 = 2lx$. Now, y tends to zero as x tends to zero. We thus have

$$f(x, \sqrt{2lx}) = 4(l - p)(l - q)x^2.$$

This quantity is positive or negative at our choice, depending on whether l is outside or inside the interval (p, q). Hence, in every neighbourhood of the values $(0, 0)$ of x and y, the function f assumes positive and negative values, or values greater than and less than $f(0, 0) = 0$. Thus f is neither maximal nor minimal.

5 AN EXAMPLE OF A FUNCTION WHOSE PARTIAL DERIVATIVES DO NOT COMMUTE

The function

$$(x, y) = xy \frac{x^2 - y^2}{x^2 + y^2}, \qquad \text{for } (x, y) \neq (0, 0),$$
$$f(0, 0) = 0$$

is a continuous function of the variables x, y and has continuous first derivatives

$$f_x(x, y) = y \frac{x^2 - y^2}{x^2 + y^2} + 4y \frac{x^2 y^2}{(x^2 + y^2)^2},$$

$$f_y(x, y) = x \frac{x^2 - y^2}{x^2 + y^2} - 4x \frac{x^2 y^2}{(x^2 + y^2)^2}$$

such that

$$f_x(0, 0) = f_y(0, 0) = 0.$$

But $f_{xy}(0, 0) = -1$ and $f_{yx}(0, 0) = 1$. We thus see that interchanging the order of differentiation is not permissible. In this case the second derivatives are discontinuous.

6 MAXIMA AND MINIMA OF FUNCTIONS OF SEVERAL VARIABLES

133. We say that $u = f(x, y, z, ...)$ becomes a maximum, or a minimum, for the values $x_0, y_0, z_0, ...$ of the variables if, for some appropriate neighbourhood of these values, we always have

$$f(x_0, y_0, z_0, ...) \geqslant f(x, y, z, ...)$$

or $\quad f(x_0, y_0, z_0, ...) \leqslant f(x, y, z, ...).$

Consider the function of x alone

$$f(x, y_0, z_0, ...).$$

If u becomes maximum or minimum for $x = x_0, y = y_0, ...$, this function becomes the same for $x = x_0$. Hence its derivative $f_x(x_0, y_0, z_0, ...)$, if it exists, is zero. We may reason analogously about the other variables and deduce:

If $u = f(x, y, z, ...)$ becomes maximum or minimum for the values $x_0, y_0, z_0, ...$, the partial derivatives of u corresponding to these values are zero.

134. Let

$$x = x_0 + ht, \qquad y = y_0 + kt, \qquad z = z_0 + lt, ...$$

and let

$$F(t) = f(x, y, z, ...).$$

If u becomes maximum or minimum for the values $x_0, y_0, z_0, ...$, the same happens for $F(t)$ at $t = 0$.

Now, if $f(x, y, z, ...)$ has continuous first partial derivatives, the function $F(t)$ has the finite derivative

$$F'(t) = f_x(x, y, z, ...)h + f_y(x, y, z, ...)k + f_z(x, y, z, ...)l + ...,$$

and hence we must have

$$F'(0) = f_x(x_0, y_0, ...)h + f_y(x_0, y_0, ...)k + ... = 0$$

whatever the values of $h, k, ...$; and as long as this condition is satisfied we must have

$$f_x(x_0, y_0, ...) = 0, \qquad f_y(x_0, y_0, ...) = 0, ...$$

as we have just found.

Suppose next that f admits continuous partial derivatives of the second order. We have

$$F''(t) = f_{xx}(x, y, ...)h^2 + f_{yy}(x, y, ...)k^2 + ... + 2f_{xy}(x, y, ...)hk + ...;$$

and if u becomes maximum or minimum for the values considered, the same happens for $F(t)$ at $t = 0$. Hence it is necessary for a maximum that

$$F''(0) \leqslant 0,$$

and for a minimum that

$$F''(0) \geqslant 0,$$

whatever the values of $h, k, ...$, or:

In order that the function u have a maximum, or a minimum, for the values $x_0, y_0, ...$ of the variables which make the first derivatives zero, it is necessary that the homogeneous function of second degree in $h, k, ...$

$$F''(0) = f_{xx}(x_0, y_0, ...)h^2 + f_{yy}(x_0, y_0, ...)k^2 + ...$$
$$+ 2f_{xy}(x_0, y_0, ...)hk + ...$$

assume, respectively, positive or negative values, whatever the values of $h, k, ...$

135. *If for $x = x_0, y = y_0, ...$ all the partial derivatives of order less than n of the function $u = f(x, y, ...)$ are zero, and if, on expanding $f(x_0 + h, y_0 + k, ...)$ by Taylor's formula, the term which contains $h, k, ...$ homogeneously of degree n is an indefinite form, u is neither a maximum nor a minimum for the values $x_0, y_0, ...$ of the variables. If, instead, this term is a positive definite form, u is a minimum, and if a negative definite form, u is a maximum.*

In fact, applying theorems 4 and 5 of no. 130 to the function

$$v = f(x, y, ...) - f(x_0, y_0, ...),$$

if the term considered is an indefinite form, in every neighbourhood of $x_0, y_0, ...$, v assumes positive and negative values, and $f(x, y, ...)$ assumes values greater than and less than $f(x_0, y_0, ...)$. Therefore the function u is neither maximum nor minimum. If, instead, it is a positive definite form, in a certain neighbourhood of $x_0, y_0, ...$, we will have $v > 0$, or

$$f(x, y, ...) > f(x_0, y_0, ...),$$

and the function u is a minimum for $x = x_0, y = y_0, ...$ We could reason in an analogous way if the term considered were negative definite.

136. Here is a criterion for recognizing whether a given homogeneous function of second degree $\psi(h, k, l, ...)$ is a positive definite form, or if,

whatever the values attributed to h, k, ..., it assumes positive values, and is never zero, unless all the variables are zero at the same time.

If ψ depends on the single variable h, we will have $\psi = Ah^2$, and it will be positive, and never zero except when h becomes zero, if $A > 0$.

If ψ depends on the two variables h and k, we will have

$$\psi = Ah^2 + 2Bhk + Ck^2;$$

and if it is a positive definite form, letting $k = 0$ and h be not equal to zero, ψ assumes the value Ah^2, which must be positive and non-zero, whence we must have

$$A > 0.$$

Also, ψ can be put into the form

$$\psi = (1/A)[(Ah + Bk)^2 + (AC - B^2)k^2],$$

and if we choose h so that $Ah + Bk = 0$, ψ assumes the positive and non-zero value $(1/A)(AC - B^2)k^2$; and in order for that to happen it is necessary that

$$AC - B^2 > 0.$$

The conditions $A > 0$ and $AC - B^2 > 0$, which are necessary in order that ψ be a positive definite quadratic form, are also sufficient. Indeed, if k is non-zero, we have

$$(AC - B^2)k^2 > 0, \qquad (Ah + Bk)^2 \geqslant 0,$$

whence their sum is positive and $\psi > 0$. If k is zero, h will not be zero, and hence ψ reduces to Ah^2, a positive quantity.

In general, if ψ depends on several variables h, k, l, ..., we can write

$$\psi = Ah^2 + 2Bh + C,$$

where A is a constant, B is a homogeneous function of the first degree in k, l, ..., and C is a homogeneous function of second degree in the same quantities. If k, l, ... become zero as h becomes zero, then B and C become zero, and ψ assumes the value Ah^2, which is positive by hypothesis, whence

$$A > 0.$$

The form ψ may be written

$$\psi = (1/A)[(Ah + B)^2 + (AC - B^2)],$$

where $AC - B^2$ is a homogeneous function of second degree in k, l, ... If we give these variables any values whatever which are not all zero, and h a value such that $Ah + B = 0$, ψ assumes the value $(1/A)(AC - B^2)$, which is positive and non-zero, and hence the expression $AC - B^2$ must be positive and non-zero. Or, letting $\psi_1(k, l, ...) = AC - B^2$, we have that ψ_1 is a homogeneous function of $k, l, ...$ which is always positive and never zero, except when all the variables are zero.

Hence the conditions which are necessary in order that ψ be a positive definite form are: (1) $A > 0$, and (2) $AC - B^2$ is a positive definite form in the variables $k, l, ...$

These conditions are sufficient. Indeed, if we give to h an arbitrary value, and to $k, l, ...$ arbitrary values not all zero, of the two terms into which ψ is decomposed, the first is positive or zero, and the second positive, whence $\psi > 0$. If, on the other hand, we let zero be the value of all the variables $k, l, ...$, h will not be zero, and ψ will have the value Ah^2, which is positive.

In this way, to recognize if a quadratic form is positive definite we are led to recognizing if another quadratic form in one variable less has that property. Continuing thus we are led to the case of one or two variables, which has already been studied.

A form ψ is negative definite if $-\psi$ is positive definite.

7 A NEW DEFINITION OF THE DEFINITE INTEGRAL

The preceding theorems suggest a new definition of the definite integral which, while not entirely equivalent to that previously given, coincides with it in the most common cases.

Let $f(x)$ be a function of x defined on an interval (a, b). Suppose that the values of $f(x)$ have a least upper bound and a greatest lower bound on the interval (a, b), and hence on every one of its subintervals.

Let $x = a, x_1, x_2, ..., x_n = b$ be arbitrarily chosen increasing values of x and let $l(\alpha, \beta)$ and $\lambda(\alpha, \beta)$ be the l.u.b. and g.l.b. of $f(x)$ on the interval (α, β). Now consider the sums

$$s_1 = \sum (x_{r+1} - x_r)l(x_r, x_{r+1}),$$
$$s_2 = \sum (x_{r+1} - x_r)\lambda(x_r, x_{r+1}),$$

where r varies from 0 to $n - 1$.

It is easy to see that every sum s_1 is greater than every sum s_2, and hence there exists a g.l.b. S_1 of the values of s_1, and an l.u.b. S_2 of the values of s_2, and $S_1 \geqslant S_2$. If $S_1 = S_2$, the common value will be that unique quantity less than all the sums s_1 and greater than all the sums s_2. In such a case we denote this common value by

$$\int_a^b f(x)dx,$$

and we shall say (in this section) that the function $f(x)$ is integrable on the interval (a, b).

III
On the integrability of
first-order differential equations (1886)*

Peano proves here for the first time the existence of a solution of the first-order differential equation $y' = f(x, y)$ on the sole condition that $f(x, y)$ be continuous. In 1890 he proved the analogous theorem for systems of ordinary differential equations, using there an entirely different method. Here the proof is elementary, if not entirely rigorous.

The proofs given heretofore of the existence of the integrals of differential equations leave something to be desired in regard to simplicity. The purpose of this note is to give an elementary proof of the existence of the integrals of the equation $dy/dx = f(x, y)$, supposing only the continuity of the function $f(x, y)$.

THEOREM

If $f(x, y)$ is a continuous function of x and y for all values of the variables which we shall consider, and if a and b are two arbitrary values of x and y, then a value A greater than a can be determined such that:

1. An infinite number of functions y_1 of x, defined on the interval (a, A), may be formed which for $x = a$ have the value b and satisfy the inequality

$$dy_1/dx > f(x, y).$$

2. An infinite number of functions y_2 of x may be formed which for $x = a$ have the value b and satisfy the inequality

$$dy_2/dx < f(x, y).$$

3. The functions y_1 have a greatest lower bound (g.l.b.) Y_1, which is a function of x, defined on the interval (a, A), which for $x = a$ has the value b, and which satisfies the equation

$$dY_1/dx = f(x, Y_1).$$

* 'Sull'integrabilità delle equazioni differenziali di primo ordine,' *Atti Accad. sci. Torino*, 21 (1885–6), 677–85 [9].

4. The least upper bound (l.u.b.) of the functions y_2 is a function Y_2 of x, defined on the interval (a, A), which for $x = a$ has the value b, and which satisfies the equation

$$dY_2/dx = f(x, Y_2).$$

5. Every function y of x which for $x = a$ has the value b and satisfies the equation

$$dy/dx = f(x, y)$$

is, in the interval (a, A), between Y_1 and Y_2:

$$Y_1 \geqslant y \geqslant Y_2.$$

6. If $f(x, y)$ has its partial derivative with respect to y less than any preassigned quantity for all values considered of the variables, all the functions y of x which for $x = a$ assume the value b and which satisfy the differential equation $dy/dx = f(x, y)$ are identical.

PROOF

Let p' be a quantity greater than $f(a, b)$ and consider the linear function

$$y' = b + p'(x - a).$$

This, for $x = a$, has the value b, and the difference

$$dy'/dx - f(x, y')$$

is a continuous function of x, which for $x = a$ has the value $p' - f(a, b) > 0$. Hence we may determine a value $a' > a$, such that, for every value of x in the interval (a, a'), we have $dy'/dx - f(x, y') > 0$, or $dy'/dx > f(x, y')$.

Let b' be the value of y' for $x = a'$ (i.e., $b' = b + p'(a' - a)$). Let p'' be a quantity greater than $f(a', b')$, and consider the function

$$y'' = b' + p''(x - a'),$$

which for $x = a'$ has the value b'. The difference $dy''/dx - f(x, y'')$ is a continuous function of x, which for $x = a'$ has the value $p'' - f(a', b') > 0$. Hence it will also be positive for all the values of x between a' and a certain value $a'' > a'$, or, in the interval (a', a''), we have

$$dy''/dx > f(x, y'').$$

Let b'' be the value of y'' for $x = a''$. Let p''' be a quantity greater than $f(a'', b'')$. Then the function

$$y''' = b'' + p'''(x - a''),$$

for $x = a''$, has the value b'', and in an interval (a'', a''') satisfies the inequality

$$dy'''/dx > f(x, y''').$$

Continuing thus, repeat this operation n times. We will have a sequence of successive intervals

$$(a, a'), (a', a''), (a'', a'''), ..., (a^{(n-1)}, a^{(n)})$$

and a sequence of functions

$$y', y'', y''', ..., y^{(n)}$$

such that the value of the first, for $x = a$, is b, and the value of each at the end of its interval is equal to the value of the next function at the beginning of the next interval. Each one of the functions satisfies, in its own interval, the inequality

$$dy/dx > f(x, y).$$

Let y_1 be the function of x (formed by a succession of linear functions) which in the successive intervals considered coincides respectively with the functions $y', y'', y''', ..., y^{(n)}$. Denote $a^{(n)}$ by A_1. We may conclude that y_1 is a continuous function of x, defined on the interval (a, A_1), which for $x = a$ has the value b, and which satisfies the inequality

$$dy_1/dx > f(x, y_1).$$

By analogous reasoning, changing the signs $>$ into $<$, we may show that a function y_2 may be formed, defined on an interval (a, A_2), which for $x = a$ has the value b, and which satisfies the inequality

$$dy_2/dx < f(x, y_2).$$

Hence, denoting by A the smaller of the values A_1 and A_2, we have formed, on the interval (a, A), functions y_1 and y_2 which satisfy all the conditions of the first and second parts of the theorem, and since we may select in an infinite number of ways the quantities $p', p'', ..., a', a'', ...$, and their analogues for y_2, we conclude that the number of functions y_1 and y_2 is infinite.

Before passing on to the other parts of the theorem, it is convenient to consider the following propositions:

I. If two functions y_1 and y_2 satisfy, respectively, the inequalities

$$dy_1/dx > f(x, y_1), \qquad dy_2/dx \leqslant f(x, y_2),$$
$$\text{or} \quad dy_1/dx \geqslant f(x, y_1), \qquad dy_2/dx < f(x, y_2),$$

and if for a special value x_0 of x they are equal, then the difference $y_1 - y_2$ will go from negative to positive as x goes from values less than x_0 to values greater than x_0.

In fact, for $x = x_0$ we obviously have

$$dy_1/dx > dy_2/dx \quad \text{or} \quad d(y_1 - y_2)/dx > 0$$

so that the difference $y_1 - y_2$ is increasing for $x = x_0$, and since it becomes zero for $x = x_0$, it will go from negative to positive.

II. If two functions y_1 and y_2 satisfy the preceding conditions, and if for $x = x_0$ we have $y_1 \geqslant y_2$, then for every value of $x > x_0$ we will have $y_1 > y_2$.

For, consider its denial. Then the difference $y_1 - y_2$ will be zero or negative for some value of $x > x_0$. Supposing that it is zero, let x_1 be the smallest value of x for which it is zero. Then the difference $y_1 - y_2$, which is not zero anywhere in the interval (x_0, x_1), will remain positive because it is positive for $x = x_0$ (if $y_1 > y_2$) or it becomes positive for $x > x_0$ (if it is zero for $x = x_0$), according to the preceding proposition. Now this is absurd, because if for $x = x_1$ the functions y_1 and y_2 are equal, for $x < x_1$, the difference $y_1 - y_2$ must be negative, by virtue of the preceding proposition. Therefore the difference $y_1 - y_2$ cannot be zero for any value of $x > x_0$, and hence cannot change signs and become negative.

Given this, the functions y_1 and y_2, which satisfy the first and second conditions of the theorem, i.e., which for $x = a$ have the value b, and which satisfy the inequalities

$$dy_1/dx > f(x, y_1), \qquad dy_2/dx < f(x, y_2),$$

will be such that for every value of x in the interval (a, A)

$$y_1 > y_2.$$

Hence, x having been given any value whatever in the interval considered, the infinite number of values which y_1 may have are all greater than the infinite number of values which y_2 may have. For this reason there exists a g.l.b. Y_1 of the values of y_1, and an l.u.b. Y_2 of the values y_2, and

$$Y_1 \geqslant Y_2.$$

Thus Y_1 and Y_2 are two functions of x, defined on the interval (a, A), which for $x = a$ have the value b, in common with all the y_1 and y_2. I assert that each one of them satisfies the proposed differential equation.

Denote Y_1 by $F(x)$. Let x be given a particular value x_0, and for brevity denote $f[x_0, F(x_0)]$ by m. Now let ε be an arbitrarily small positive quantity, and consider the function

$$\varphi(x) = F(x_0) + (x - x_0)(m - \varepsilon).$$

For $x = x_0$, this has the value $F(x_0)$, and, for the same value of the variable, it satisfies the inequality $dy/dx < f(x, y)$. Hence it satisfies this inequality for all values of x between x_0 and a certain value $x_1 > x_0$. Now, each function y_1, which satisfies the conditions of the first part of the theorem, has for $x = x_0$ a value greater than $F(x_0) = \varphi(x_0)$, because $F(x_0)$ is the g.l.b. of the values of the functions y_1. Moreover, it satisfies the inequality $dy_1/dx > f(x, y_1)$. Hence, by a known proposition, we will have, for each value of x in the interval (x_0, x_1), $y_1 > \varphi(x)$, so that $F(x)$, i.e., the g.l.b. of the values of y_1, will not be less than $\varphi(x)$, i.e., $F(x) \geq \varphi(x)$, or, substituting for $\varphi(x)$,

$$F(x) \geq F(x_0) + (x - x_0)(m - \varepsilon),$$

which may be written

$$\frac{F(x) - F(x_0)}{x - x_0} \geq m - \varepsilon.$$

On the other hand, let $\varepsilon > 0$ again be arbitrarily fixed. Then the quantity

$$H = m + \varepsilon - f[x, F(x_0)] + \alpha + (m + \varepsilon)(x - x_0)$$

is a continuous function of α and x, which for $\alpha = 0$ and $x = x_0$ is reduced to ε, a positive quantity. Hence we may determine $\rho > 0$ and $x_1 > x_0$ in such a way that, for every value of $\alpha < \rho$ and for every value of x in the interval (x_0, x_1), we have $H > 0$. Now, since $F(x_0)$ is the g.l.b. of the values which are assumed, for $x = x_0$, by the functions y_1, we may determine one of these functions which assumes, for $x = x_0$, a value $F(x_0) + \alpha$, where $\alpha < \rho$. Consider now the function $\psi(x)$ which on the interval (a, x_0) coincides with the function y_1 just considered, and which in the interval (x_0, x_1) has the value

$$\psi(x) = F(x_0) + \alpha + (m + \varepsilon)(x - x_0).$$

We have $d\psi/dx - f(x, \psi) = H > 0$. Hence, in the interval (x_0, x_1) we have $d\psi/dx > f(x, \psi)$. Therefore the function $\psi(x)$ satisfies all the conditions of part 1 of the theorem. But $F(x)$ is the g.l.b. of the functions which satisfy these conditions, so that in the interval (x_0, x_1) we have $F(x) < \psi(x)$, or

$$F(x) < F(x_0) + \alpha + (m + \varepsilon)(x - x_0).$$

This inequality is satisfied no matter what α is, which we may take to be arbitrarily small. Hence

$$F(x) \leqslant F(x_0) + (m + \varepsilon)(x - x_0),$$

which we may write

$$\frac{F(x) - F(x_0)}{x - x_0} \leqslant m + \varepsilon.$$

The inequalities we have found,

$$m - \varepsilon \leqslant \frac{F(x) - F(x_0)}{x - x_0} \leqslant m + \varepsilon,$$

were proved for $x > x_0$, but because of their symmetry in x and x_0, we may remove this hypothesis. They say in fact that

$$F'(x_0) = m,$$

or, substituting for m, that the differential equation

$$F'(x) = f[x, F(x)]$$

is satisfied for every value x_0 of x in the interval (a, A).

In an analogous way we may prove that Y_2 satisfies the same differential equation. (For that matter, if we let $y = -z$, then the functions y_1, Y_1, y_2, Y_2 are changed respectively into y_2, Y_2, y_1, Y_1.) Thus we have proved the third and fourth parts of the theorem.

Now let y_1, y_2, and y be three functions of x, which for $x = a$ have the value b, and which satisfy the conditions

$$dy_1/dx > f(x, y_1), \qquad dy_2/dx < f(x, y_2), \qquad dy/dx = f(x, y).$$

By a known proposition we have in the interval (a, A)

$$y_1 > y > y_2.$$

Hence Y_1, the g.l.b. of the functions y_1, and Y_2, the l.u.b. of the y_2, satisfy the conditions

$$Y_1 \geqslant y \geqslant Y_2,$$

which is the fifth part of the theorem.

Admitting solely the continuity of $f(x, y)$ it is not possible to deduce any consequence beyond the existence of the two functions Y_1 and Y_2, which for $x = a$ have the value b, which satisfy the equation $dy/dx = f(x, y)$ and which have between them all functions that satisfy the same equation and for $x = a$ have the value b. But, if we make the hypothesis of part 6 of the theorem, then all these functions coincide.

In fact, from the equations

$$dY_1/dx = f(x, Y_1), \qquad dY_2/dx = f(x, Y_2),$$

we infer that

$$d(Y_1 - Y_2)/dx = f(x, Y_1) - f(x, Y_2),$$
$$\text{or} \quad d(Y_1 - Y_2)/dx = (Y_1 - Y_2)f_y(x, y),$$

where y is a value between Y_1 and Y_2. Let us suppose now that for every value of x between a and A, and for all values of y between Y_1 and Y_2, we have $f_y(x, y) < M$, where M is a finite constant. Since $Y_1 - Y_2 \geqslant 0$, we deduce that

$$d(Y_1 - Y_2)/dx \leqslant (Y_1 - Y_2)M.$$

We may integrate this inequality by moving everything to the left-hand side and multiplying by e^{-Mx}, a positive quantity. Thus $d[e^{-Mx}(Y_1 - Y_2)]/dx \leqslant 0$. Therefore the function $e^{-Mx}(Y_1 - Y_2)$ is never increasing on the interval (a, A). It is zero for $x = a$, because for this value of x, Y_1 and Y_2 have the value b. It cannot become negative because $e^{-Mx} > 0$ and $Y_1 \geqslant Y_2$. Hence it is zero for every value of x, or, in other words, the functions Y_1 and Y_2, and all functions between them, coincide throughout the interval (a, b).[1]

---

1 Cauchy proved for the first time the existence of one and only one function y of x which satisfies the equation $dy/dx = f(x, y)$ and which for $x = a$ has a given value b, supposing however that $f(x, y)$ is a monogenic function of the variables. This proof was simplified by C. Briot and J. Bouquet (*J. Ecole polytechnique*, 21 [1856], 133–98). New proofs of the same proposition, without introducing the consideration of monogenic functions, but with some restrictions on the nature of the function $f(x, y)$, were given by Messrs Lipschitz [*Ann. mat. pura appl.*, (2) 2 (1868), 288–302], Hoüel, Gilbert, et al. Mr Vito Volterra (*Giornale di matematiche*, vol. 19 [1880]) generalized these results, leaving in doubt, however, the truth of the theorem supposing only the continuity of $f(x, y)$. In the present note we also have the answer to this question. [Many proofs of this theorem have since been given. One along the lines suggested by Peano is: Helmut Grunsky, 'Ein konstruktiver Beweis für die Lösbarkeit der Differentialgleichung $f' = f(x, y)$ bei stetigem $f(x, y)$,' *Jber. Deutsch. math. Verein.*, 63 (1960), 78–84.]

IV

Integration by series of
linear differential equations (1887)*

This article contains Peano's first use of the method of successive approximations (or successive integrations) for the solution of linear differential equations. A slightly modified version, in French, was published the following year in the Mathematische Annalen. *Peano always believed that he was the first to discover this method, and he tried several times to vindicate his priority over Emile Picard, who, Peano believed, began using it in 1891. In fact, Picard had already given an example of successive approximations in 1888 and, at that time, gave credit for the method to H.A. Schwartz.*

1

The purpose of the present note is to prove the following theorem. Let

$$dx_1/dt = \alpha_{11}x_1 + \alpha_{12}x_2 + \ldots + \alpha_{1n}x_n,$$
$$dx_2/dt = \alpha_{21}x_1 + \alpha_{22}x_2 + \ldots + \alpha_{2n}x_n,$$
$$\cdot \; \cdot \; \cdot \; \cdot \; \cdot \; \cdot \; \cdot \; \cdot \; \cdot \; \cdot \; \cdot \; \cdot \; \cdot \; \cdot \; \cdot \; \cdot \; \cdot \; \cdot \; \cdot$$
$$dx_n/dt = \alpha_{n1}x_1 + \alpha_{n2}x_2 + \ldots + \alpha_{nn}x_n$$

be n homogeneous linear differential equations in n functions $x_1, x_2, \ldots, x_n$ of the variable t, in which the coefficients α_{ij} are functions of t which are continuous on an interval $p \leqslant t \leqslant q$.

Substitute in the second members of the proposed differential equations, in place of the $x_1, \ldots, x_n$, n arbitrary constants $a_1, a_2, \ldots, a_n$, and integrate from t_0 to t, t_0 and t being in the interval (p, q). We obtain n functions of t, which will be denoted by $a_1', a_2', \ldots, a_n'$.

Substitute in the second members of the given differential equations, in place of $x_1, \ldots, x_n$, respectively $a_1', \ldots, a_n'$, and integrate from t_0 to t. We obtain n new functions of t, which will be denoted by $a_1'', a_2'', \ldots, a_n''$.

Substitute in the second members of the given equations, in place of the x the a'', and integrate from t_0 to t. We obtain the functions $a_1''', \ldots, a_n'''$.

* 'Integrazione per serie delle equazioni differenziali lineari,' *Atti Accad. sci. Torino*, 22 (1886–7), 437–46 [10].

Continuing this process, we obtain the series

$$a_1 + a_1' + a_1'' + ...,$$
$$a_2 + a_2' + a_2'' + ...,$$
$$. . . . .$$
$$a_n + a_n' + a_n'' + ...,$$

which are convergent throughout the interval (p, q). Their sums, which we shall denote by $x_1, x_2, ..., x_n$, are functions of t which satisfy the proposed differential equations, and which, for $t = t_0$, assume the arbitrarily chosen values $a_1, a_2, ..., a_n$.

We thus have the general integral of the given equations expressed by means of convergent series provided the given functions α_{ij} are continuous. The terms of these series are obtained from the α_{ij} by the sole operations of addition, multiplication, and integration. Moreover, the convergence of these series, comparable to that of the development of e^x, is in general sufficiently rapid to be useful for calculations.

2

In order to prove the preceding proposition, and others analogous to it, it is almost essential to introduce some notation, based on number complexes of any order, to simplify our expressions.

We define *number complex of order n* to be a set of n real numbers. The complex formed from the numbers $a_1, a_2, ..., a_n$ will be indicated by $[a_1, a_2, ..., a_n]$. When it is not necessary to display the real numbers which make up the complex, it will be indicated by a single letter $\mathbf{a} = [a_1, a_2, ..., a_n]$.

Two complexes $\mathbf{a} = [a_1, ..., a_n]$ and $\mathbf{b} = [b_1, ..., b_n]$ of order n are said to be equal if the numbers which compose them are equal in the same order. Hence, the equality $\mathbf{a} = \mathbf{b}$ between two complexes implies the n equalities between real numbers $a_1 = b_1, ..., a_n = b_n$.

If $\mathbf{a} = [a_1, ..., a_n]$ and $\mathbf{b} = [b_1, ..., b_n]$ are two complexes of order n, we define their sum to be

$$\mathbf{a} + \mathbf{b} = [a_1 + b_1, ..., a_n + b_n].$$

It follows that $\mathbf{a} + \mathbf{b} = \mathbf{b} + \mathbf{a}$ and $\mathbf{a} + (\mathbf{b} + \mathbf{c}) = (\mathbf{a} + \mathbf{b}) + \mathbf{c}$, or addition of number complexes enjoys the commutative and associative properties of real numbers.

If $\mathbf{a} = [a_1, ..., a_n]$ is a complex of order n, and k is a real number, we define their product to be

$$k\mathbf{a} = [ka_1, ka_2, ..., ka_n].$$

It follows that $(k + k')\mathbf{a} = k\mathbf{a} + k'\mathbf{a}$ and $k(\mathbf{a} + \mathbf{b}) = k\mathbf{a} + k\mathbf{b}$, or the multiplication of a real number and a complex is distributive with respect to both factors.

With these definitions we have determined the meaning of the expression

$$k\mathbf{a} + k'\mathbf{a}' + k''\mathbf{a}'' + ...,$$

where k, k', k'', ... are real numbers and $\mathbf{a}$, $\mathbf{a}'$, $\mathbf{a}''$, ... are complexes of the same order.

If we let $\mathbf{i}_1 = [1, 0, 0, ..., 0]$, $\mathbf{i}_2 = [0, 1, 0, ..., 0]$, ..., $\mathbf{i}_n = [0, 0, ..., 0, 1]$, then every number complex $\mathbf{x} = [x_1, x_2, ..., x_n]$ can be expressed by a sum

$$\mathbf{x} = x_1\mathbf{i}_1 + x_2\mathbf{i}_2 + ... + x_n\mathbf{i}_n.$$

If, to the expression $\mathbf{a} - \mathbf{b}$ we attribute the meaning $\mathbf{a} + (-1)\mathbf{b}$, and to $\mathbf{a}/k$, where k is real, the meaning $(1/k)\mathbf{a}$, then the difference of two complexes and the quotient of a complex by a real number have also been defined, and we have $(\mathbf{a} - \mathbf{b}) + \mathbf{b} = \mathbf{a}$ and $k(\mathbf{a}/k) = \mathbf{a}$.

We shall define the modulus of a complex $\mathbf{x} = [x_1, x_2, ..., x_n]$ by the equality

$$\mathrm{mod}\, x = \sqrt{x_1{}^2 + x_2{}^2 + ... + x_n{}^2}.$$

It may easily be shown that

$$\mathrm{mod}(\mathbf{a} + \mathbf{b}) \leqslant \mathrm{mod}\,\mathbf{a} + \mathrm{mod}\,\mathbf{b},$$
$$\mathrm{mod}(k\mathbf{a}) = (\mathrm{mod}\,k)(\mathrm{mod}\,\mathbf{a}),$$

where $\mathbf{a}$ and $\mathbf{b}$ are complexes, k is a real number, and $\mathrm{mod}\,k$ is the absolute value of k.

We shall say that the variable complex $\mathbf{x} = [x_1, ..., x_n]$ has as limit the complex $\mathbf{a} = [a_1, ..., a_n]$, if $\lim x_1 = a_1$, $\lim x_2 = a_2$, ..., $\lim x_n = a_n$. We deduce that, if $\lim \mathbf{x} = \mathbf{a}$, then $\lim \mathrm{mod}(\mathbf{x} - \mathbf{a}) = 0$, and conversely.

Having defined sum and limit of complexes, we may extend the definition of convergence to series of complexes. It can be shown that a series of complexes is convergent if the series formed by their moduli is convergent.

If $\mathbf{x} = [x_1, x_2, ..., x_n]$ is a complex function of the real variable t, we can extend to it the definitions of derivative and integral. Thus

$$d\mathbf{x}/dt = [dx_1/dt, dx_2/dt, ..., dx_n/dt],$$

$$\int_{t_0}^{t} \mathbf{x}dt = \left[\int_{t_0}^{t} x_1 dt, \int_{t_0}^{t} x_2 dt, ..., \int_{t_0}^{t} x_n dt \right]$$

(or we may assume these equalities as the definitions of derivative and integral of a complex).

It can be shown that, if $t_0 < t_1$,

$$\mathrm{mod}\left(\int_{t_0}^{t_1} \mathbf{x}dt\right) < \int_{t_0}^{t_1} (\mathrm{mod}\ \mathbf{x})dt.$$

3

We shall designate as *linear transformation of a complex* the operation by which to each complex $\mathbf{x} = [x_1, x_2, ..., x_n]$ is made to correspond a new complex

$$[\alpha_{11}x_1 + \alpha_{12}x_2 + ... + \alpha_{1n}x_n, \alpha_{21}x_1 + \alpha_{22}x_2 + ... + \alpha_{2n}x_n,$$
$$..., \alpha_{n1}x_1 + \alpha_{n2}x_2 + ... + \alpha_{nn}x_n],$$

such that the numbers which make it up are homogeneous linear functions of the numbers which compose the complex $\mathbf{x}$. The linear transformation considered depends on the n^2 coefficients α_{ij}, and will be indicated by the schema

$$\left\{\begin{matrix} \alpha_{11} & \alpha_{12} & \cdots & \alpha_{1n} \\ \alpha_{21} & \alpha_{22} & \cdots & \alpha_{2n} \\ \cdots & \cdots & \cdots & \cdots \\ \alpha_{n1} & \alpha_{n2} & \cdots & \alpha_{nn} \end{matrix}\right\}.$$

When it is not necessary to display the coefficients of the transformation, it will be indicated by a single letter α, β, ... If $\mathbf{x}$ is a complex and α a transformation, by $\alpha\mathbf{x}$ we shall understand the new complex obtained by carrying out on $\mathbf{x}$ the transformation α.

We may show that if $\mathbf{x}$ and $\mathbf{y}$ are number complexes, α a linear transformation, then

$$\alpha(\mathbf{x} + \mathbf{y}) = \alpha\mathbf{x} + \alpha\mathbf{y}.$$

Conversely, if $\alpha\mathbf{x}$ is a complex function of the complex $\mathbf{x}$ such that $\alpha(\mathbf{x} + \mathbf{y}) = \alpha\mathbf{x} + \alpha\mathbf{y}$, and if, as $\mathbf{x}$ tends toward $\mathbf{x}_0$, $\lim \alpha\mathbf{x} = \alpha\mathbf{x}_0$, then the complex $\alpha\mathbf{x}$ can be obtained by operating on $\mathbf{x}$ with a linear transformation.

Two linear transformations α and β are said to be equal if, for an arbitrary complex $\mathbf{x}$, we have $\alpha\mathbf{x} = \beta\mathbf{x}$. It follows that if $\alpha = \beta$, then each of the n^2 coefficients in α is equal to the corresponding one in β. For a transformation α to be equal to a number k it is necessary and sufficient that all the elements of α which lie on the principal diagonal be equal to k, and the others be zero.

If α and β are two linear transformations, the number complex $\alpha\mathbf{x} + \beta\mathbf{x}$ can be obtained by carrying out on $\mathbf{x}$ a new linear transformation, which we indicate by $\alpha + \beta$; thus by definition

$$(\alpha + \beta)\mathbf{x} = \alpha\mathbf{x} + \beta\mathbf{x}.$$

If

$$\alpha = \left\{ \begin{matrix} \alpha_{11} & \cdots & \alpha_{1n} \\ \cdots\cdots\cdots \\ \alpha_{n1} & \cdots & \alpha_{nn} \end{matrix} \right\}, \qquad \beta = \left\{ \begin{matrix} \beta_{11} & \cdots & \beta_{1n} \\ \cdots\cdots\cdots \\ \beta_{n1} & \cdots & \beta_{nn} \end{matrix} \right\},$$

then

$$\alpha + \beta = \left\{ \begin{matrix} \alpha_{11} + \beta_{11} & \cdots & \alpha_{1n} + \beta_{1n} \\ \cdots\cdots\cdots\cdots\cdots\cdots \\ \alpha_{n1} + \beta_{n1} & \cdots & \alpha_{nn} + \beta_{nn} \end{matrix} \right\}.$$

If on the complex x we first operate with the transformation α, and then with the transformation β, we obtain the complex $\beta\alpha x$, which could also be obtained by operating on x with the single transformation $\beta\alpha$. If α and β represent the transformations indicated by the preceding schema, then $\beta\alpha$ will be represented by the schema

$$\beta\alpha = \left\{ \begin{matrix} \beta_{11}\alpha_{11} + \beta_{12}\alpha_{21} + \ldots + \beta_{1n}\alpha_{n1} \\ \qquad \beta_{11}\alpha_{12} + \beta_{12}\alpha_{22} + \ldots + \beta_{1n}\alpha_{n2} & \cdots \\ \beta_{21}\alpha_{11} + \beta_{22}\alpha_{21} + \ldots + \beta_{1n}\alpha_{n1} \\ \qquad \beta_{21}\alpha_{12} + \beta_{22}\alpha_{22} + \ldots + \beta_{2n}\alpha_{n2} & \cdots \\ \cdots\cdots\cdots\cdots\cdots\cdots\cdots\cdots\cdots\cdots \end{matrix} \right\}.$$

The transformation $\beta\alpha$ is said to be the product of the two transformations α and β.

Thus we have defined what we mean by any expression whatever formed by real numbers and linear transformations combined according to the operations of addition and multiplication; it represents a linear transformation. For these operations the following identities hold:

$$\alpha + \beta = \beta + \alpha, \qquad \alpha + (\beta + \gamma) = (\alpha + \beta) + \gamma,$$
$$\alpha(\beta + \gamma) = \alpha\beta + \alpha\gamma, \qquad (\alpha + \beta)\gamma = \alpha\gamma + \beta\gamma.$$

If α is a transformation, k a real number, we have $k\alpha = \alpha k$; but, if α and β are two arbitrary transformations, it is no longer true in general that $\alpha\beta = \beta\alpha$.

Let x be a number complex of order n, and α be a linear transformation. The ratio $[\mathrm{mod}(\alpha x)]^2/(\mathrm{mod}\ x)^2$ is the ratio of two homogeneous forms of second degree in $x_1, x_2, \ldots, x_n$, which are the real numbers which compose x. The denominator $x_1^2 + x_2^2 + \ldots + x_n^2$ is a positive definite form; the numerator is positive or zero. Hence, by a known proposition of calculus, this ratio has a maximum for a certain set of values of $x_1, x_2, \ldots, x_n$, i.e., for a certain complex x. This maximum is positive or zero; its square root will be given the name *modulus of the transformation* α. Thus

$$(\mathrm{mod}\ \alpha x)^2/(\mathrm{mod}\ x)^2 \leqslant (\mathrm{mod}\ \alpha)^2,$$

or $\quad \mathrm{mod}(\alpha x) \leqslant (\mathrm{mod}\ \alpha)(\mathrm{mod}\ x).$

This inequality can serve as a definition of mod α, if we understand that the first member is never greater than the second, and the first is equal to the second for one value of $\mathbf{x}$.

It can be shown that

$$\mod(\alpha + \beta) \leqslant \mod \alpha + \mod \beta,$$
$$\mod \alpha\beta \leqslant (\mod \alpha)(\mod \beta).$$

The modulus of a transformation is a continuous function of the coefficients of the transformation, which is bounded if these are bounded.

We shall say that a variable transformation α has α_0 as limit if, for an arbitrary complex $\mathbf{x}$, we have $\lim \alpha\mathbf{x} = \alpha_0\mathbf{x}$. It follows that if $\lim \alpha = \alpha_0$, then all the coefficients of the schema of α must have as limit the corresponding ones of α_0. Having defined the limit of a transformation, we may define the convergence of a series whose terms are linear transformations. Thus, a series whose terms are transformations is convergent if the series formed of the moduli of the transformations is convergent.

If α is a transformation, a function of a real variable t, we can extend to it the definitions of derivative and integral. Thus, if $t_0 < t_1$, then

$$\mod \int_{t_0}^{t_1} \alpha dt \leqslant \int_{t_0}^{t_1} (\mod \alpha)dt,$$

and if

$$\alpha = \begin{Bmatrix} \alpha_{11} & \cdots & \alpha_{1n} \\ \cdots & \cdots & \cdots \\ \alpha_{n1} & \cdots & \alpha_{nn} \end{Bmatrix},$$

then

$$\frac{d\alpha}{dt} = \begin{Bmatrix} \dfrac{d\alpha_{11}}{dt} & \cdots & \dfrac{d\alpha_{1n}}{dt} \\ \cdots & \cdots & \cdots \\ \dfrac{d\alpha_{n1}}{dt} & \cdots & \dfrac{d\alpha_{nn}}{dt} \end{Bmatrix}, \qquad \int \alpha dt = \begin{Bmatrix} \int\alpha_{11}dt & \cdots & \int\alpha_{1n}dt \\ \cdots & \cdots & \cdots \\ \int\alpha_{n1}dt & \cdots & \int\alpha_{nn}dt \end{Bmatrix}.$$

4

Using these notations, let

$$\mathbf{x} = [x_1, x_2, ..., x_n], \qquad \alpha = \begin{Bmatrix} \alpha_{11} & \cdots & \alpha_{1n} \\ \cdots & \cdots & \cdots \\ \alpha_{n1} & \cdots & \alpha_{nn} \end{Bmatrix}.$$

The differential equations proposed are represented by the single equation

$$d\mathbf{x}/dt = \alpha\mathbf{x}.$$

Let $\mathbf{a} = [a_1, ..., a_n]$ be an arbitrarily chosen constant complex. Let

$$\mathbf{a}' = \int_{t_0}^{t} \alpha \mathbf{a} \, dt, \qquad \mathbf{a}'' = \int_{t_0}^{t} \alpha \mathbf{a}' \, dt, \qquad \mathbf{a}''' = \int_{t_0}^{t} \alpha \mathbf{a}'' \, dt.$$

Then the real numbers which constitute a', a'', ... are precisely the numbers a_1', ..., a_n', a_1'', ..., a_n'', ... introduced in the statement of the theorem.

Since the functions α_{ij} are continuous and bounded on the interval (p, q), the same will be true of mod α. Hence, letting M denote the maximum value of mod α on this interval, we have

$$\mathrm{mod}\, \mathbf{a}' < M \, \mathrm{mod} \frac{t - t_0}{1} \, \mathrm{mod}\, \mathbf{a}, \quad \mathrm{mod}\, \mathbf{a}'' < \frac{M \, \mathrm{mod}(t - t_0)^2}{2!} \, \mathrm{mod}\, \mathbf{a}, ...$$

$$\mathrm{mod}\, \mathbf{a}^{(p)} < \frac{[M \, \mathrm{mod}(t - t_0)]^p}{p!} \, \mathrm{mod}\, \mathbf{a}, ...$$

Now, the series

$$\mathrm{mod}\, \mathbf{a} + \frac{M \, \mathrm{mod}(t - t_0)}{1} \, \mathrm{mod}\, \mathbf{a} + \frac{[M \, \mathrm{mod}(t - t_0)]^2}{2!} \, \mathrm{mod}\, \mathbf{a} + ...$$

is uniformly convergent throughout the interval (p, q), and has as its sum

$$e^{M \, \mathrm{mod}(t - t_0)} \, \mathrm{mod}\, \mathbf{a}.$$

Hence, the series

(A) $\quad \mathbf{a} + \mathbf{a}' + \mathbf{a}'' + \mathbf{a}''' + ...$

whose moduli are less than the terms of the preceding series is also convergent, and hence the real series

$$a_1 + a_1' + a_1'' + ...,$$
$$a_2 + a_2' + a_2'' + ...,$$
$$. \; . \; . \; . \; . \; . \; . \; . \; . \; . \; .$$

are also convergent.

The derivatives of the terms of series (A) are

$$0, \alpha \mathbf{a}, \alpha \mathbf{a}', \alpha \mathbf{a}'', ...$$

which form the series (A) multiplied by α. Hence this series is also convergent and uniformly convergent. Therefore, if we let

$$\mathbf{x} = \mathbf{a} + \mathbf{a}' + \mathbf{a}'' + ...,$$

we have

$$d\mathbf{x}/dt = \alpha \mathbf{a} + \alpha \mathbf{a}' + \alpha \mathbf{a}'' + ...$$

or

$$dx/dt = \alpha x,$$

i.e., **x** effectively satisfies the proposed differential equation. If we then let $t = t_0$, we have $\mathbf{a}' = 0$, $\mathbf{a}'' = 0$, ..., and hence **x** = **a**. We have thus proved the theorem.

Substituting in the development $\mathbf{x} = \mathbf{a} + \mathbf{a}' + \mathbf{a}'' + ...$ the values of $\mathbf{a}'$, $\mathbf{a}''$, ..., the series which gives **x** can be put into the form

$$\mathbf{x} = (1 + \int \alpha dt + \int \alpha \int \alpha dt^2 + \int \alpha \int \alpha \int \alpha dt^3 + ...)\mathbf{a},$$

in which each integral is taken from t_0 to t.

5

If the α are independent of t, i.e., the proposed differential equations have constant coefficients, letting $t_0 = 0$, we have

$$\mathbf{x} = \left(1 + \alpha t + \frac{(\alpha t)^2}{2!} + \frac{(\alpha t)^3}{3!} + ...\right) \mathbf{a};$$

and if we agree to represent by e^α, where α is an arbitrary complex, the sum of the series $1 + \alpha + \alpha^2/2! + ...$, the integral of the proposed differential equation becomes $\mathbf{x} = e^{\alpha t}\mathbf{a}$.

A homogeneous linear differential equation of order n containing only one function is, as is known, reducible to a system of linear equations of first order. If we apply to these the preceding series we get the development given by Sturm (*Cours d'analyse* [2nd ed., 1880], p. 614), by Caqué (*Journal de Liouville*, 1864), and by Fuchs (*Annali di matematica*, 1870).

The integral of the non-homogeneous linear differential equations

$$dx_1/dt = \alpha_{11}x_1 + ... + \alpha_{1n}x + p_1,$$

$$\cdot \cdot \cdot \cdot \cdot \cdot \cdot \cdot \cdot \cdot \cdot \cdot$$

$$dx_n/dt = \alpha_{n1}x_1 + ... + \alpha_{nn}x_n + p_n$$

can be obtained, as is known, from the integral of the same equations in which 0 is substituted for p. If, besides the preceding conventions, we let

$$\mathbf{p} = [p_1, ..., p_n],$$

the proposed differential equations are reduced to

$$dx/dt = \alpha x + \mathbf{p}.$$

Indicating by ε the sum of the series already considered,

$$1 + \int \alpha dt + \int \alpha \int \alpha dt^2 + ...,$$

in which the integrals are taken from t_0 to t, and by ε^{-1} the analogous series in which the integrals are taken from t to t_0, the integral of the proposed differential equation, which for $t = t_0$ assumes the value a, is given by

$$\mathbf{x} = \varepsilon \mathbf{a} + \varepsilon \int_{t_0}^{t} \varepsilon^{-1} \mathbf{p} dt.$$

The transformations ε and ε^{-1} satisfy the condition $\varepsilon \varepsilon^{-1} = 1$.

V
Geometrical magnitudes (1887)*

In 1885 Peano began teaching the course on 'geometrical applications of the infinitesimal calculus' at the University of Turin, and in 1887 he published a text with this title. The following four excerpts are taken from Chapter v, 'Geometrical Magnitudes.' They are:

1. *Point sets (pp. 152–5).*
2. *Areas (pp. 155–7).*
3. *Volumes (pp. 158–60).*
4. *Integrals extended to point sets (pp. 185–7).*

1 POINT SETS

1. There are collections of geometrical magnitudes which are such that, for two magnitudes of the same collection, there are only two possibilities: (1) The two magnitudes can be superposed or they can be decomposed into parts which are superposable two by two. In this case the two magnitudes are said to be equal. (2) The two magnitudes can be decomposed into parts in such a way that each part of the second is equal to one of the parts of the first, but not vice versa. In this case the two magnitudes are said to be unequal, and the first greater than the second.

Such is the case for the lengths of rectilinear segments, for areas of planar regions bounded by straight lines, for volumes of prisms or solids decomposable into prisms, and for several other types of geometrical magnitudes. All these are said to be *principal*. But there are other magnitudes for which it may happen that, on comparing two, neither of the two possibilities is true. For these magnitudes it is necessary to define carefully what is meant by the equality of two magnitudes, and by the measure of such a magnitude.

2. We shall define *point set* (or sometimes *figure*) to be any set of points, whether limited in number or not. Thus a finite number of points, or the

---

* Excerpts from *Applicazioni geometriche del calcolo infinitesimale* (Turin: Bocca, 1887) [11].

points of a line, a surface, or a solid, are point sets. A point set is said to be *linear* if all the points lie on one straight line; it is said to be *planar* if all the points lie in a plane. We begin our study with linear sets.

Let A be a linear set. We shall say that a point P is an *interior point* of the set A if it is possible to determine a length ρ such that all the points of the straight line whose distance from P is less than ρ belong to the set A. We shall say that a point is an *exterior point* of the set A if it is possible to determine a length ρ such that all the points of the straight line whose distance from the point P is less than ρ do not belong to A. A point which is neither interior nor exterior to A is said to be a *boundary point* of A Hence, if P is a boundary point of A, and a length ρ is arbitrarily chosen, then there will always be points of the straight line whose distance from P is less than ρ and which belong to A, and points of the line whose distance from P is less than ρ and which do not belong to A The boundary points of A may or may not belong to A. They form a new set which is called the *boundary set* of A.

It is known that every point P of a line may be made to correspond to its abscissa, i.e. the number which measures its distance from a fixed point of the line, taking due account of its sign. Conversely, to every number there corresponds, on the line, one and only one point having that number as its abscissa. Hence to every A corresponds a set of numbers, or *number set*. Conversely, to every number set corresponds a point set on the line. Because of this one-to-one correspondence we shall be able, wherever it is convenient, to consider number sets instead of point sets.

As an example, consider on the line the points whose abscissas are greater than 0 and less than 1. We have a linear point set; every point of the set is an interior point of it, the boundary points are the points with abscissas 0 and 1, and every other point is an exterior point of the set.

Consider now the points of the line whose abscissas are rational numbers greater than 0 and less than 1. This set has no interior points; the points whose abscissas are ≥ 0, and ≤ 1, are boundary points; all others are exterior points.

If a point set contains some but not all the points of a line, it will certainly have boundary points. Indeed, suppose that P is a point of a given set A, and Q a point which does not belong to A. Let p and q be the abscissas of the points P and Q, with, say, $p < q$. The points of the set A whose abscissas are less than q will have a least upper bound (l.u.b.) which is not less than p, because p is precisely the abscissa of one such point, and not greater than q. Let r be this l.u.b. and R the point of the line having r as abscissa. I assert that R is a boundary point of A. Indeed, let a length ρ be arbitrarily chosen. Then there exists some point of the set A whose abscissa is greater than

$r - \rho$, and hence such that its distance from R is less than ρ. But each point whose abscissa is between r and the smaller of the two quantities $r + \rho$ and q, the distance of these points from R also being less than ρ, does not belong to the set.

The points of a line which lie between two given points, whether these two points are included or not, form a set which is said to be a *linear segment*. Its length is a principal magnitude; every set formed by a finite number of segments also has a length comparable to that of a linear segment.

Consider now an arbitrary set of points on a line. We can imagine point sets formed by a finite number of segments of which the given set forms a part; and we can imagine also point sets formed by a finite number of segments which make up part of the given set. Each of these point sets has a length, and the length of each of the first sets is greater than the length of each of the second sets.

If the greatest lower bound (g.l.b.) of the lengths of the first sets coincides with the l.u.b. of the lengths of the second sets, then the common value of these two is said to be the *length of the given linear set*.

It may happen, however, that the l.u.b. and the g.l.b. just mentioned are not equal, and hence that the g.l.b. of the first lengths is greater than the l.u.b. of the second lengths. In this case we shall say that the proposed set does not have a length comparable to that of a linear segment. The g.l.b. of the first lengths might be called the *exterior length* of the given set, and the l.u.b. of the second lengths the *interior length*. It could happen that no sets formed of segments which contain the given set exist. We then say that the exterior length of the set is infinite. It could also happen that no segments contained in the given set exist. We then say that the interior length is zero. Thus, of the two examples given above, the first set is a segment having length equal to 1, and the second set does not have length comparable with that of a segment, its exterior length being equal to 1 and its interior length being zero.

If from a set formed of a finite number of segments, and containing A in its interior, is removed a set also formed of a finite number of segments contained in the interior of A, we will have a set likewise formed of a finite number of segments, whose length is equal to the difference between the length of the first and of the second sets, and which contains in its interior all the boundary points of A. Now in order that A have a length comparable to that of a segment it is necessary and sufficient that the difference between the lengths of the first sets and those of the second sets can be made as small as we please. Hence it is necessary and sufficient that we can construct a set formed of a finite number of segments, containing in its interior all the boundary points of A, and of magnitude as small as we please. In every case we see that the difference between the exterior length and the interior length of a set A is equal to the exterior length of the boundary set of A.

2 AREAS

The statements made about linear sets may be repeated for point sets which lie in the same plane. We shall say that a point P is *interior* to the plane A if it is possible to determine a length ρ such that all points of the plane which are a distance from P less than ρ pertain to A. A point is said to be *exterior* to the set A if it is interior to the set formed of the points not belonging to A. A point which is neither interior nor exterior is said to be a *boundary point*. The set formed by the boundary points of A is said to be the *boundary set*, or *boundary* of A.

If the set A contains some points of the plane, without containing all, it will have boundary points. Indeed, if P and Q are two points, the one pertaining to the set A and the other not, then consider the set formed of the points of the set A which lie on the line PQ. It will have, from what we have just shown, at least one boundary point belonging to the segment PQ, and this will be a boundary point of the set A.

Now consider an arbitrary planar set A. We can imagine planar regions, bounded by straight lines, which contain the set A in their interior, and we can imagine planar regions, also bounded by straight lines, contained in the interior of the given set. If, as happens in the most common cases, the g.l.b. of the areas of the first regions coincides with the l.u.b. of the areas of the second regions, to the common value of these two we shall give the name *area of the given set*. But it can happen that the g.l.b. and the l.u.b. just mentioned are not equal. We then give the name *exterior area* of the given figure to the g.l.b. of the areas of the polygonal regions which contain in their interior the given figure, and the name *interior area* of the figure to the l.u.b. of the areas of the polygonal regions contained in the interior of it.

If from a set bounded by straight lines, containing in its interior the set A, we remove a set also bounded by straight lines and contained in A, we shall obtain a (strip-like) set bounded by straight lines and containing in its interior the boundary of A. The area of this set is the difference between the areas of the first two. Hence we can assure ourselves that the g.l.b. of the first areas coincides with the l.u.b. of the second areas if their difference can be made as small as we please. That is to say, in order that a planar set have an area comparable to a polygonal area it is necessary and sufficient that a planar set can be formed which is bounded by straight lines, containing in its interior all the boundary points of the given set, and whose area is as small as we please. It can also happen that no polygon, of finite area, containing in its interior the given set exists. We then say that the exterior area of this set is infinite. If no polygon contained in the interior of the given set exists, then we say that its interior area is zero.

It is known (e.g., in elementary geometry) that if the given figure is a circle, then the l.u.b. of the areas of the interior polygons coincides with the g.l.b. of the areas of the polygons which contain the circle, and that hence the circle has an area comparable with polygonal areas. If we consider the set formed by the points of the plane whose distance from a fixed point O is rational (with respect to a length equal to 1) and less than 1, we shall have a planar set whose interior area is zero, and whose exterior area is equal to the area of the circle of radius 1.

3 VOLUMES

Consider a point set in space. A point is said to be *interior* to the set if there exists a sphere with centre the point considered, such that all points in its interior belong to the set. If instead there exists a sphere with centre the point considered such that none of its points belong to the set, the point is said to be *exterior*. A point which is neither interior nor exterior is said to be a *boundary point*. The boundary points form a set which is said to be the *boundary set*, or *boundary* of the given set.

Given a set A, we can imagine solids formed of prisms (which we also call prismatic solids) which contain A in their interior, and also prismatic solids interior to A. If the g.l.b. of the volumes of the first solids coincides with the l.u.b. of the volumes of the second solids, to their common value we shall give the name *volume* of the given set, and we also say that the given set has a volume comparable to the volumes of prisms.

But if the g.l.b. and the l.u.b. just mentioned do not coincide, we give the name *exterior volume* of the set to the g.l.b. of the volumes of the prismatic solids containing the given set, and the name *interior volume* to the l.u.b. of the volumes contained in the interior of the given set. It could happen that no prismatic solid containing the given set exists. We then say that the exterior volume of the set is infinite. If no solid is contained in the set, then we say that its interior volume is zero.

If from a solid composed of prisms, containing the set A in its interior, we remove an analogous solid contained in A, we shall have a new solid which contains the boundary points of A, and whose volume is the difference between the volumes of the two preceding solids. If this difference can be made as small as we please, then the g.l.b. of the volumes of the first solids is equal to the l.u.b. of the volumes of the second solids, and conversely.

Hence, in order that a set have a volume comparable with prismatic volumes, it is necessary and sufficient that a prismatic solid can be formed whose volume is arbitrarily small, and which contains in its interior the boundary of the given set.

It is known from elementary geometry how the volume of the most common solids, such as tetrahedra, other polyhedra, and spheres, can be shown to be comparable with that of prismatic solids.

We next observe, apropos both volumes and areas, that if a magnitude a is the l.u.b. of a system of magnitudes b, and if each magnitude b is the l.u.b. of certain magnitudes c, then the magnitude a is also the l.u.b. of the magnitudes c; and if a is the g.l.b. of a system of magnitudes b, and each magnitude b is the g.l.b. of other magnitudes c, then a will be the g.l.b. of the c. Hence, seeing that the area of a circle, or of a figure bounded by straight lines and by circular arcs, is at the same time the l.u.b. of the areas of the polygons interior to it, and the g.l.b. of the polygons which contain it, we deduce that the interior area of an arbitrary figure is also the l.u.b. of all the planar areas contained in it, and bounded by straight lines or by circular arcs (or in general by curves which enclose regions comparable in area to polygons). A similar remark may be made about exterior areas.

Further, since it is easy to see that the area of a polygon is the l.u.b. of the areas of figures interior to it, and composed of rectangles having two sides parallel to a fixed straight line, and is the g.l.b. of the areas of analogous figures, which contain the given polygon, it likewise follows that the interior area of an arbitrary set is also the l.u.b. of the areas of figures composed of rectangles having a pair of sides parallel to a fixed line, and interior to the given set, and the exterior area of the same set is the g.l.b. of the areas of analogous figures containing the given set.

Analogously, we may conclude that the interior volume of an arbitrary set is also the l.u.b. of the volumes of solids limited by planes or by spherical surfaces, or cylindrical or conical surfaces, etc. (which solids are more general than those used in the definition), and that it is also the l.u.b. of the volumes of solids formed by rectangular prisms whose altitudes are parallel to a fixed line (solids less general than those used in the definition).

4 INTEGRALS EXTENDED TO POINT SETS

21. Let a magnitude x be a distributive function of a point set which assumes only positive values. [On page 167 the following definition is given: 'A magnitude is said to be a *distributive function* of a point set if the value of this magnitude corresponding to the set is the sum of the values of it corresponding to the parts into which the given set may be partitioned.'] To each

point of the set considered let there correspond a number ρ, which can vary with the point. We define the *integral of ρ with respect to x extended to the set A* to be a magnitude such that: (1) it is always greater than the result obtained by partitioning the set A into parts, in any way whatever, multiplying the value of x corresponding to each of these parts by a number less than every value assumed by ρ in this partial set, and then summing these products; (2) it is less than the sum of the products of the values of x, corresponding to the parts of A, by the respective numbers greater than those assumed by ρ on the same parts; (3) and it is the only magnitude which enjoys these properties.

We shall indicate the integral of ρ with respect to x extended to the set A by writing $\int_A \rho dx$. Thus by $\int_A \rho dx$ we understand a magnitude (homogeneous with x) such that, the set A having been arbitrarily partitioned, say into $A = A_1 + A_2 + \dots + A_n$, indicating by $x(A_1)$, $x(A_2)$, ..., $x(A_n)$ the corresponding values of x, and letting ρ_1', ρ_2', ..., ρ_n' and ρ_1'', ρ_2'', ..., ρ_n'' be numbers such that the ρ' are less and the ρ'' greater than the values assumed by ρ on these partial sets, the following inequalities will always be satisfied:

$$\int_A \rho dx > \rho_1'x(A_1) + \rho_2'x(A_2) + \dots + \rho_n'x(A_n),$$
$$\int_A \rho dx < \rho_1''x(A_1) + \rho_2''x(A_2) + \dots + \rho_n''x(A_n),$$

however the set A is partitioned and however the values of the ρ_i' and the ρ_i'' are taken, so long as ρ_i' is less than the values assumed by ρ on the set A_i, and ρ_i'' is greater than the same values; and such that it is the only magnitude which satisfies these conditions

22. Suppose that the values of ρ corresponding to the points of A are bounded and consider the sums:

$$s' = \rho_1'x(A_1) + \rho_2'x(A_2) + \dots + \rho_n'x(A_n)$$

and

$$s'' = \rho_1''x(A_1) + \rho_2''x(A_2) + \dots + \rho_n''x(A_n),$$

which depend on the law by which A is divided into parts, and on the choice of the numbers ρ_i' and ρ_i''.

Each sum s' is always less than every sum s'', whether these correspond to the same partition of A or to different partitions. This is evident if s' and s'' correspond to the same partition of A, since ρ_i' is then less than every value of ρ on the set A_i and ρ_i'' is greater than all of them. Hence $\rho_i' < \rho_i''$, and multiplying by $x(A_i)$, a positive quantity, we have $\rho_i'x(A_i) < \rho_i''x(A_i)$, and then by summing, $s' < s''$. If s' and s'' correspond to different partitions of A, consider that partition of A which results from their superposition.

Substituting in s' and s'' for the terms $x(A_i)$ the sum of the values of x corresponding to the parts into which A_i is partitioned, s' and s'' become the sum of the values of x corresponding to the new partition of A, multiplied respectively by the numbers less than and greater than the values assumed by ρ on these sets, and hence we always have $s' < s''$.

Therefore, the quantities s' will have an l.u.b. and the s'' a g.l.b., and the l.u.b. of the s' will be less than or equal to the g.l.b. of the s''.

If the l.u.b. of the s' is equal to the g.l.b. of the s'', their common value will be $\int_A \rho \, dx$, because this will be a quantity always greater than the values of s', less than the values of s'', and the only quantity between the values of s' and s''. But if the l.u.b. of the s' is less than the g.l.b. of the s'', then these two quantities and every quantity between them are greater than the s' and less than the s'', and we can no longer speak of integral in the sense just defined. We shall give these quantities names, however; we shall say that the l.u.b. of the values of s' is the *lower integral* of ρ with respect to x and indicate it by $\underline{\int}_A \rho \, dx$, and we shall say that the g.l.b. of the values of s'' is the *upper integral* of ρ with respect to x and indicate it by $\overline{\int}_A \rho \, dx$.

It could also happen that no finite values ρ' and ρ'' exist between which all the values of ρ lie. In this case, again, we may not speak of an integral, properly so-called, but we may still have one of the two integrals, lower or upper, or we may have neither.

VI

The geometrical calculus according to the Ausdehnungslehre of H. Grassmann, preceded by the operations of deductive logic (1888)*

We include two chapters from this monograph of 1888. The first is the introductory chapter, 'The Operations of Deductive Logic.' (It has almost no connection with the following chapters.) In it we find Peano's earliest treatment of deductive logic, based on his study of works of G. Boole, E. Schröder, C.S. Peirce, and others. Here Peano notes the equivalence of the calculus of sets and the calculus of propositions. Several of the symbols used in this selection were not used in later publications.

The second selection is Chapter I, 'Geometrical Formations.' This and the remaining chapters of the book result from Peano's reworking of Grassmann's Ausdehnungslehre. *Peano made no claim to originality for the ideas contained in it, but there can be no doubt that the extreme clarity of his presentation, in contrast with the notorious difficulty of reading Grassmann's work, helped to spread Grassmann's ideas and make them more popular.*

1 THE OPERATIONS OF DEDUCTIVE LOGIC

Notation

1

Let there be given a system of entities, and let A, B, ... be classes of this system.

For example we may consider the system of all real numbers, and some classes of this system will be: the rational numbers, the integers, the multiples of a given number, the numbers which can be roots of algebraic equations with rational coefficients, and so on. If a is a number, the class formed by the numbers greater than a is defined, and we shall indicate it in these examples by $(>a)$; the class of numbers less than a is likewise defined, and we shall indicate it by $(<a)$.

* Excerpts from *Calcolo geometrico secondo l'Ausdehnungslehre di H. Grassmann, preceduto dalle operazioni della logica deduttiva* (Turin: Bocca, 1888) [14].

We introduce the following notation:

1. By the expression $A = B$ we affirm the identity of the two classes A and B, that is to say, every entity in A is also in B, and vice versa. The sign $=$ is read *equals*; the proposition $A = B$ is said to be a *logical equation*; A and B are *members* of this equation.

EXAMPLES

'even number' $=$ 'multiple of 2'

'rational number' $=$ 'number that can be developed as a finite continued fraction'

2. By the expression $A \cap B \cap C \cap ...$, or $ABC...$, we mean the largest class contained in the classes $A, B, C, ...$ or the class formed by all the entities which are at the same time in A and B and C, etc. The sign $\cap$ is read *and*; the operation indicated by the sign $\cap$ is logical *conjunction*. We shall also call it *logical multiplication*, and say that the classes $A, B, ...$ are *factors* of the *product AB...*

EXAMPLES

'multiple of 6' $=$ 'multiple of 2' $\cap$ 'multiple of 3'

'$(>1) \cap (<2)$' $=$ 'the system of numbers between 1 and 2'

3. By the expression $A \cup B \cup C \cup ...$ we mean the smallest class which contains the classes $A, B, C, ...$, or the class formed of the entities which are in A or B or C, etc. The sign $\cup$ is read *or*; the operation indicated by the sign $\cup$ is called logical *disjunction*. We shall also call it *logical addition*, and say that the classes $A, B, ...$ are *terms* of the *sum $A \cup B \cup ...$*

EXAMPLES

'rational number' $=$ 'integer' $\cup$ 'fraction'

'$(<1) \cup (>2)$' $=$ 'numbers not between 1 and 2, and different from 1 and 2'

4. By the expression $-A$, or $\bar{A}$, we mean the class formed by all the entities not belonging to the class A. The sign $-$ is read *not*; the operation indicated by the sign $-$ is called *negation*.

EXAMPLES

$-$'rational' $=$ 'irrational'

$-(>a) =$ 'class of numbers less than or equal to a'

5. In order for the preceding operations to always have meaning, it is necessary to consider as a class the set of all entities of the system, indicated by the sign $\oslash$, and read *all*. It is also necessary to consider as a class the lack of any entity, indicated by the sign $\bigcirc$, and read *empty*. Hence the expression '$A = \bigcirc$' represents the proposition 'there is no A'; '$AB = \bigcirc$' expresses

the negative universal proposition 'no A is a B'; '$A \cup B = \oslash$' says 'everything is an A or a B.'

EXAMPLES

$(>1) \cup (<2) = \oslash$

$(<1) \cap (>2) = \bigcirc$

'integers' $\cup$ 'fractions' $\cup$ 'irrationals' $= \oslash$

'rationals' $\cap$ $-$'integers' $\cap$ 'roots of an integral rational algebraic equation with integral coefficients, and in which the coefficient of the term of highest degree is unity' $= \bigcirc$

6. The expression $A\bar{B} = \bigcirc$ says that entities do not exist which are in A and at the same time not in B; it is substantially equivalent to the universal affirmative proposition 'every A is a B.' Even though the preceding expression for indicating this proposition is already quite simple, nonetheless for greater convenience we shall indicate it also by the expression

$$A < B, \quad \text{or } B > A,$$

which can be read 'every A is a B,' or 'the class B contains A.' The signs $<$ and $>$ can also be read *less than* and *greater than*.

EXAMPLES

'integers' $<$ 'rationals'   'rationals' $>$ 'fractions'

$(<1) < (<2)$ $(>1) > (>2)$

Identities

2

The following identities are evident:

(1) $AB = BA$,

(2) $A(BC) = ABC$,

(3) $AA = A$,

(4) $A(B \cup C) = AB \cup AC$,

(5) $A\oslash = A$,

(6) $A\bigcirc = \bigcirc$,

(7) $-(-A) = A$,

(8) $-(AB) = (-A) \cup (-B)$,

(9) $A \cap -A = \bigcirc$,

(10) $-\bigcirc = \oslash$,

(1') $A \cup B = B \cup A$,

(2') $A \cup (B \cup C) = A \cup B \cup C$,

(3') $A \cup A = A$,

(4') $A \cup BC = (A \cup B)(A \cup C)$,

(5') $A \cup \bigcirc = A$,

(6') $A \cup \oslash = \oslash$,

(8') $-(A \cup B) = (-A) \cap (-B)$,

(9') $A \cup -A = \oslash$,

(10') $-\oslash = \bigcirc$.

The identities (1), (1'), (2), (2'), and (4) express the commutative and associative properties of the operations $\cap$ and $\cup$, and the distributive property of the operation $\cap$ with respect to $\cup$. These properties furnish sufficient reason for giving the names multiplication and addition to these operations. Formulas (5) and (5′) say that $\bigcirc$ is the identity of the operation $\cup$, and $\oslash$ that of $\cap$. The properties expressed by the identities (3) and (3′) have no corresponding ones in algebra. Formula (4′), of which we shall make little use, expresses the distributive property of the operation $\cup$ with respect to $\cap$; so that the operations $\cup$ and $\cap$ are commutative, associative, each distributive with respect to the other; they are the only operations in the operational calculus now known which enjoy these properties. The operations $\cap$ and $\cup$ do not have inverse operations in the strict sense of this term; in some way the operation $-$ supplies this lack; formula (7) says that two negations make an affirmation; (8) says that the negative of a product is the sum of the negatives of the factors, a property analogous to that of logarithms; (8′) says that the negative of a sum is the product of the negatives of the terms. Formula (8′) may also be written $A \cup B = -[(-A)(-B)]$; hence, having introduced the signs $\cap$ and $-$ we may do without the sign $\cup$. Formulas (10) and (10′) say that it is sufficient to introduce only one of the two signs $\bigcirc$ and $\oslash$.

The preceding formulas are all evident, and every argument makes continuous use of them. Some of them must be held as axioms, and the others are their consequences; in fact the remaining formulas can be obtained from (1), (2), (3), (4), (5), (7), (8), (9), (10), by substituting in these, for certain expressions, others which are equivalent by virtue of these particular formulas.

Thus, in order to derive (8′), we have the series of equalities

$$(-A) \cap (-B) \underset{(7)}{=} -\{-[(-A) \cap (-B)]\}$$
$$\underset{(8)}{=} -[(--A) \cup (--B)] \underset{(7)}{=} -(A \cup B),$$

by virtue of the identities indicated by the numbers placed under the sign $=$; the equality of the first and the last members gives precisely the formula to be proved.

We also have

$$A \bigcirc \underset{(9)}{=} A(A \cap -A) \underset{(2)}{=} A \cap A \cap -A \underset{(3)}{=} A \cap -A \underset{(9)}{=} \bigcirc,$$

and the equality of the first and the last members gives precisely formula (6).

The formulas with accents are then deduced from those without by changing $A, B, \ldots$ into $\bar{A}, \bar{B}, \ldots$ and taking the negatives of both members.

We also have the identities

(11) $A \cup AB = A$,   (11′) $A(A \cup B) = A$.

For

$$A \cup AB = A \oslash \cup AB = A(\oslash \cup B) = A \oslash = A,$$
$$\quad (5) \qquad\qquad (4) \qquad\qquad (1'),\,(6') \quad (5)$$

and the equality of the first and the last members gives formula (11). The proof of (11') is analogous.

Since $A < B$ is identical with $A\bar{B} = \bigcirc$, the formulas

(12) $AB < A,$ \qquad\qquad (12') $A \cup B > A,$

(13) $\bigcirc < A,$ \qquad\qquad (13') $\oslash > A,$

are no more than a new way of writing the identities

$$AB\bar{A} = \bigcirc, \qquad A\bar{A}B = \bigcirc, \qquad \bigcirc\bar{A} = \bigcirc, \qquad A\,\overline{\oslash} = \bigcirc.$$

3

An expression obtained by operating on a class X, and on other classes taken to be fixed, with the logical signs $\cap$, $\cup$, $-$ is said to be a function of X, and is indicated by $f(X)$. By virtue of the preceding identities it may be put in many different forms.

We shall say that a logical expression as a function of X is in a *separated* form if it has the form

(a) $f(X) = PX \cup Q\bar{X},$

where P and Q are classes independent of X. Every logical function of X can always be put in a separated form, and in one way only. In fact, any class A independent of X can be put in the separated form

$$A = AX \cup A\bar{X}.$$

The class X can be put in the separated form

$$X = \oslash\, X \cup \bigcirc\, \bar{X}.$$

The sum of two separated expressions

$$(PX \cup Q\bar{X}) \cup (P'X \cup Q'\bar{X}) = (P \cup P')X \cup (Q \cup Q')\bar{X}$$

remains expressed in a separated form; the product of two separated forms becomes, after some calculation,

$$(PX \cup Q\bar{X})(P'X \cup Q'\bar{X}) = PP'X \cup QQ'\bar{X},$$

a separated form; the negation of a separated form

$$-(PX \cup Q\bar{X}) = \bar{P}X \cup \bar{Q}\bar{X}$$

also remains expressed in a separated form. Hence, every logical function of X, which is obtained by operating a finite number of times on constant classes and on X with the operations $\cup$, $\cap$, $-$, may be written in a separated form.

It is easy to see the significance of the coefficients P and Q. If in (a) we let $X = \varnothing$, and then $X = \bigcirc$, we get

$$f(\varnothing) = P, \qquad f(\bigcirc) = Q;$$

hence P and Q have a well-determined value. Substituting in (a), we have the relation

(14) $f(X) = f(\varnothing)X \cup f(\bigcirc)\overline{X}$,

which presents an analogy with Taylor's formula.

If $f(X, Y)$ is a logical function of the two variable classes X and Y, and of other classes taken as fixed, we will have, by (14),

$$f(X, Y) = f(\varnothing, Y)X \cup f(\bigcirc, Y)\overline{X},$$

and, again applying formula (14) to the coefficients,

(15) $f(X, Y) =$
$$f(\varnothing, \varnothing)XY \cup f(\varnothing, \bigcirc)X\overline{Y} \cup f(\bigcirc, \varnothing)\overline{X}Y \cup f(\bigcirc, \bigcirc)\overline{X}\,\overline{Y}.$$

Propositions

4

A proposition may express a relation among determinate entities; it is then said to be *categorical*; considered in itself it can only be true or false. Or a proposition may contain indeterminate (variable) entities, and it is then said to be *conditional*; the condition expressed by such a proposition is in general true of certain entities, and not of others; but the condition could be true of all entities, or of none. Every affirmation is made with categorical propositions; a conditional proposition cannot be part of a categorical proposition. Thus, for example, the proposition '$x^2 - 3x + 2 = 0$' is conditional, since it contains the indeterminate x; the proposition 'the equation $x^2 - 3x + 2 = 0$ has 1 and 2 as roots' is categorical, having as subject the preceding conditional. Thus, too, '$(x + y)^2 = x^2 + 2xy + y^2$' is conditional, whereas '$(x + y)^2 = x^2 + 2xy + y^2$ is an identity' is categorical.

If the conditional proposition α contains the indeterminate entity x, by $x{:}\alpha$ we understand the class formed of all the entities for which the proposition α is true. If a proposition α contains several indeterminate entities

$x, y, \ldots$, we consider as a new entity the set of one x, of one $y, \ldots$, and we indicate it by $(x, y, \ldots)$. By the expression $(x, y, \ldots):\alpha$ we understand the class formed by all the entities $(x, y, \ldots)$ for which α is true. If the proposition α does not contain any indeterminate entities other than $x, y, \ldots$, the class $(x, y, \ldots):\alpha$ is a well-determined class, and a proposition which affirms some property of this class is categorical. If, instead, the proposition α contains, besides the entities $x, y, \ldots$, still other variable entities $u, v, \ldots$ (and is a function of these entities), the proposition which affirms a property of this class will be a conditional proposition in $u, v, \ldots$

The sign : may be read *such that*.

EXAMPLES Let $x, y, \ldots$ be numbers and $f, \varphi, \ldots$ be symbols of numerical functions; then:

The expression $x:[f(x) = 0]$ represents the class of numbers x such that $f(x)$ is zero, i.e. the roots of the equation $f(x) = 0$.

The expression $x:[f(x) = 0] \cap x:[\varphi(x) = 0]$ represents the common roots of the two equations $f(x) = 0$ and $\varphi(x) = 0$.

The expression $x:[f(x) = 0] = x:[\varphi(x) = 0]$ expresses the proposition 'the equations $f(x) = 0$ and $\varphi(x) = 0$ have the same roots.'

$x:[f(x)\varphi(x) = 0] = x:[f(x) = 0] \cup x:[\varphi(x) = 0]$ says that 'the roots of the equation $f(x)\varphi(x) = 0$ are roots of $f(x) = 0$ or roots of $\varphi(x) = 0$.'

$x:[f(x) = 0] = \oslash$ says that '$f(x) = 0$ is an identity with respect to x.'

$(x, y):[f(x, y) = 0]$ represents the set of all pairs of values of x and y such that $f(x, y) = 0$.

$x:[f(x, y) = 0]$ represents the set of values of x such that $f(x, y) = 0$; it depends on the value of y (supposing the function f is given).

$x:[f(x, y) = 0] = \oslash$ says that '$f(x, y) = 0$ is an identity with respect to x'; it is a conditional proposition in y.

$y:\{x:[f(x, y) = 0] = \oslash\}$ represents the class of values of y such that $f(x, y) = 0$ is an identity with respect to x.

$y:\{x:[f(x, y) = 0] = \oslash\} = \bigcirc$ says: 'there do not exist values of y such that $f(x, y)$ is zero for every value of x.' This proposition is categorical if $f(x, y)$ is a given function, conditional if $f(x, y)$ is not a given function.

5

In order to simplify the expression in a discussion when the system of entities $(x, y, \ldots)$ considered as variables is well defined, we shall omit from in front of the conditional the sign $(x, y, \ldots):$. We shall adopt this simplification only when all entities represented by letters are considered as variables. Hence, if $\alpha, \beta, \ldots$ are conditional propositions, we have:

$\alpha < \beta$, or $\beta > \alpha$, says that 'the class defined by the condition α is part of that defined by β,' or 'α has β as consequent,' 'β is a consequence of α,' 'if α is true, then β is true.'

α = β says that 'if α is true, then β is true, and vice versa.'

α ∩ β expresses the condition that α and β are true at the same time.

α ∪ β expresses the condition that α is true or β is true.

−α expresses the condition obtained by negating α.

○ expresses an *absurd* condition.

⊘ expresses the condition of *identity*.

If A is a class, $-[A = ○]$ says 'it is not true that the class A does not contain any entity' or 'the class A contains some entity'; the expression $-[AB = ○]$ expresses the particular affirmative proposition 'some A is a B.'

EXAMPLES In order to acquaint himself with the signs introduced, the reader may interpret in ordinary language the following propositions, in which $a, b, ..., x, y, ...$ represent real finite numbers:

$$(a < b) = (b > a), \quad -(a < b) = (a \geqslant b), \quad -(a = b) = (a \gtrless b),$$
$$(a = b) = (a + c = b + c),$$
$$(a = b) < (ac = bc), \quad (ac = bc) = (a = b) \cup (c = 0),$$
$$(ac = bc) \cap -(c = 0) < (a = b).$$
$$(a = b) < (a^2 = b^2), \quad (a^2 = b^2) = (a = b) \cup (a = -b),$$
$$(ax + b = a'x + b') = [(a - a')x = b' - b],$$
$$(x + y = a) \cap (x - y = b) = (2x = a + b) \cap (2y = a - b),$$
$$(xy = 0) = (x = 0) \cup (y = 0),$$
$$((x^2 + y^2) = 0) = (x = 0) \cap (y = 0),$$
$$(x^2 - 3x + 2 > 0) = (x < 1) \cup (x > 2),$$
$$-(x^2 - 3x + 2 > 0) = -(x < 1) \cap -(x > 2),$$
$$[(x + y)^2 = x^2 + 2xy + y^2] = ⊘, \quad (x^2 + y^2 + 1 = 0) = ○,$$
$$[x: (ax^2 + bx + c = a'x^2 + b'x + c') = ⊘] =$$
$$(a = a') \cap (b = b') \cap (c = c')$$

(i.e., in order for the equality $ax^2 + ... = a'x^2 + ...$ to be satisfied for every value of x, it is necessary and sufficient that we have at the same time $a = a', b = b', c = c'$),

$$\left\{ x: \left[\frac{1}{x(x - 1)} = \frac{a}{x - 1} + \frac{b}{x} \right] = ⊘ \right\} = (a = 1) \cap (b = -1),$$
$$[x: (x^2 + y^2 = 1) = ○] = (y < -1) \cup (y > +1)$$

(i.e., in order that the equation $x^2 + y^2 = 1$ not have real roots in x, it is necessary and sufficient that either y be less than -1, or that y be greater than 1).

Operations on propositions

6

We shall review rapidly some logical identities containing conditional propositions.

Indicating by A, B, ... certain classes, or conditional propositions defining these classes, we have the identities:

[1] $(A = B) = (B = A)$,

[2] $(A < B) = (A\bar{B} = \bigcirc)$,

[3] $(A > B) = (B < A)$,

[4] $(A = B) = (A > B) \cap (A < B)$.

The identities [1] and [3] say that: *Every logical equation is changed into an equal one by exchanging the two members, and the signs* $=$, $<$, $>$ *into* $=$, $>$, $<$.

The identities [2] and [3] express the definitions given for the signs $<$ and $>$ by using the sign $=$; [4] expresses the sign $=$ using the signs $>$ and $<$.

Further, we have the following identities, which express the uniformity of the operations $\cap$, $\cup$, $-$:

[5] $(A = B) < (AC = BC)$,

[6] $(A = B) < (A \cup C = B \cup C)$,

[7] $(A = B) < (-A = -B)$.

Applying twice the identities [5] and [6] we get

[5'] $(A = B) \cap (A' = B') < (AA' = BB')$,

[6'] $(A = B) \cap (A' = B') < (A \cup A' = B \cup B')$.

The identities

[5''] $(A < B) < (AC < BC)$,

[6''] $(A < B) < (A \cup C < B \cup C)$,

[5'''] $(A < B) \cap (A' < B') < (AA' < BB')$,

[6'''] $(A < B) \cap (A' < B') < (A \cup A' < B \cup B')$

may be deduced from the preceding by substituting for the equation $A < B$ its equivalent $A\bar{B} = \bigcirc$. Thus, from [5], we have $(A\bar{B} = \bigcirc) < (A\bar{B}C = \bigcirc)$;

now $A\bar{B}C = A\bar{B}C \cup A\bar{C}C = AC(\bar{B} \cup \bar{C}) = AC\bar{B}\bar{C}$ by identities [9], [4], [8]; hence, by substituting, $(A\bar{B} = \bigcirc) < (AC\bar{B}\bar{C} = \bigcirc)$ or $(A < B) < (AC < BC)$, which is [5''].

The preceding identities say that: *From a system of logical equations, all true, and all containing the same sign* $=$, *or* $<$, *or* $>$, *new equations, also true, are deduced by multiplying both members by the same factor or by adding the same term, or summing them member by member, or by multiplying them member by member.*

If in identity [7] we change A and B into $-A$ and $-B$ we deduce that

$$(-A = -B) < (A = B);$$

this identity along with [7] gives, by [4],

[7'] $(A = B) = (-A = -B)$.

We then have

$$(A < B) = (A\bar{B} = \bigcirc) = [\bar{B}(-\bar{A}) = \bigcirc] = [\bar{B} < \bar{A}] = (\bar{A} > \bar{B}),$$

or

[8] $(A < B) = (\bar{A} > \bar{B})$.

Identities [7'] and [8] say that: *Every logical equation is transformed into an equivalent one by taking the negatives of each of its members and changing the signs* $=$, $<$, $>$ *into* $=$, $>$, $<$.

The following identity is important:

[9] $(A \cup B = \bigcirc) = (A = \bigcirc) \cap (B = \bigcirc)$.

In fact, multiplying both its members by A, we have

$$(A \cup B = \bigcirc) < (A \cup AB = \bigcirc)$$

and, since by (11) $A \cup AB = A$, we have

(α) $(A \cup B = \bigcirc) < (A = \bigcirc)$.

Multiplying by B the two members of the proposed equation [9], we obtain analogously

(β) $(A \cup B = \bigcirc) < (B = \bigcirc)$.

Multiplying member by member (α) and (β), we deduce that

(γ) $(A \cup B = \bigcirc) < (A = \bigcirc) \cap (B = \bigcirc)$.

Summing member by member $A = \bigcirc$ and $B = \bigcirc$, we have

(δ) $\quad (A = \bigcirc) \cap (B = \bigcirc) < (A \cup B = \bigcirc)$,

and the combination of (γ) and (δ) says precisely that

$$(A \cup B = \bigcirc) = (A = \bigcirc) \cap (B = \bigcirc).$$

7

By virtue of the preceding identities a logical equation may assume several different forms.

Thus the universal affirmative proposition 'every A is a B' is indicated by the expression (§1, 6)

(a) $\quad A < B$.

Exchanging the members [3] it becomes

(b) $\quad B > A$

or 'the class B contains A.'

Multiplying both members of (a) by A we deduce that

(c) $\quad A < AB$;

and since we have identity (12) $AB < A$, from this and from (c) we deduce that

(c′) $\quad A = AB$.

Adding B to both members of (a), we get

(d) $\quad A \cup B < B$,

which, joined with identity (12′) $A \cup B > B$, gives

(d′) $\quad A \cup B = B$.

Now multiplying both members of (a) by $\bar{B}$, we will have

(e) $\quad A\bar{B} < \bigcirc$

and, by identity (13) $A\bar{B} > \bigcirc$, we have

(e′) $\quad A\bar{B} = \bigcirc$,

or 'no A is a non-B.'

Adding $\bar{A}$ to both members of (a), we will have

(f) $\oslash < \bar{A} \cup B$

and, by identity (13') $\oslash > \bar{A} \cup B$, we have

(f') $\oslash = \bar{A} \cup B$,

or 'everything is a non-A, or is a B.'

Taking the negatives [8] of both members of (a), we have, by exchanging the members,

(g) $\bar{B} < \bar{A}$,

or 'every non-B is a non-A.'

In this way, from proposition (a) we have deduced the others, (b), ..., (g), by very simple logical operations. As an exercise, the reader may deduce all the others from another of these propositions. Hence they are all equivalent, and are only different ways of expressing 'every A is a B.'

8

The four propositions

(I)   every A is a B,

(II) no A is a B,

(III) some A is a B,

(IV) some A is not a B

may be expressed, as we have just seen, by the expressions

(I) $A\bar{B} = O$,

(II) $AB = O$,

(III) $-(AB = O)$,

(IV) $-(A\bar{B} = O)$.

Expressions (I) and (II) are called *universal* by logicians; (III) and (IV), which are the negations of universal propositions, are called *particular*; (I) and (III), which contain an even number of negations, are called *affirmative*; (II) and (IV) are *negative*.

We have

$$(AB = O) = (BA = O),$$

or   'no A is a B' = 'no B is an A.'

We also have

$$-(AB = \bigcirc) = -(BA = \bigcirc),$$

or 'some A is a B' = 'some B is an A.'

These equalities constitute the so-called *inversions* of the universal negative and particular affirmative propositions.

Propositions (I) and (IV), and also (II) and (III), of which one is the negative of the other, are called *contradictory*. We have

$$(AB = \bigcirc) \cap -(AB = \bigcirc) = \bigcirc,$$
$$(A\bar{B} = \bigcirc) \cap -(A\bar{B} = \bigcirc) = \bigcirc,$$

or the coexistence of two contradictory propositions is absurd.

Propositions (I) and (II) are called *contraries*; it is stated in textbooks in logic that two contrary propositions cannot coexist. We have arrived at a somewhat different result. In fact, we have, by formula [9],

$$(AB = \bigcirc) \cap (A\bar{B} = \bigcirc) = (AB \cup A\bar{B} = \bigcirc),$$

or $\quad (AB = \bigcirc) \cap (A\bar{B} = \bigcirc) = (A = \bigcirc),$

that is to say, the coexistence of propositions (I) and (II) is equivalent to $A = \bigcirc$; multiplying this equation by $-(A = \bigcirc)$ we deduce that

[10] $\quad (AB = \bigcirc) \cap (A\bar{B} = \bigcirc) \cap -(A = \bigcirc) = \bigcirc,$

i.e., propositions (I) and (II) cannot coexist, supposing that class A is not empty. Certainly, when the logicians affirm that two contrary propositions cannot coexist, they understand that class A is not empty; but although all the rules given by the preceding formulas are true no matter what the classes which make them up, including $\bigcirc$ and $\oslash$, this is the first case in which it is necessary to suppose that one of the classes considered is not empty.

Formula [10] can also be written

[10'] $\quad (A\bar{B} = \bigcirc) \cap -(A = \bigcirc) < -(BA = \bigcirc),$

or 'if every A is a B, and if the class A is not empty, it follows that some B is an A.' Hence, given the convention of considering $\bigcirc$ and $\oslash$ as classes also, 'some B is an A' is a consequence not of the proposition 'every A is a B' alone, but of this and 'the class A is not empty.'

9

The propositions 'every A is a B' and 'every B is a C' may be written

$$A\bar{B} = \bigcirc, \quad B\bar{C} = \bigcirc.$$

Multiplying the first by $\bar{C}$, the second by A, and summing, we have

$$A\bar{C} = O$$

or 'every A is a C.' We thus have the simplest form of the syllogism

[11] $(A\bar{B} = O) \cap (B\bar{C} = O) < (A\bar{C} = O)$,

which can also be written

[11'] $(A < B) \cap (B < C) < (A < C)$,

or

[11''] $(A\bar{B} = O) \cap (B\bar{C} = O) \cap -(A\bar{C} = O) = O$,

or, changing C into $\bar{C}$,

[11'''] $(A\bar{B} = O) \cap -(AC = O) < -(BC = O)$.

This last formula may be stated as: 'if every A is a B, and some A is a C, then some B is a C.'

If in the forms [11] and [11'''] we change B into $\bar{B}$, or C into $\bar{C}$, or both at the same time, and exchange the factors of the first member, the syllogism takes various forms, some of which have been considered by logicians and called *modes and figures*. But among the modes considered by logicians some cannot be reduced to the preceding form, namely those in which from two general propositions is deduced a particular. Thus we cannot obtain the form 'every B is a C, and every B is an A; therefore some A is a C.'

The reason is easy to discover, for while the preceding formulas, and the syllogisms derived from them, are true whatever the classes introduced, even if these include O or $\oslash$, in this new form it is necessary to suppose that the class B is not empty. In fact we have, by formula [9],

$$(B\bar{C} = O) \cap (B\bar{A} = O) \cap (AC = O) = [B(\bar{C} \cup \bar{A}) \cup AC = O]$$
$$= [B \cap -(AC) \cup AC(B \cup \bar{B}) = O]$$
$$= \{B \cap [-(AC) \cup AC] \cup AC\bar{B} = O\}$$
$$= [B \cup AC\bar{B} = O] = (B = O) \cap (AC\bar{B} = O).$$

Hence, multiplying by $-(B = O)$, we have

$$(B\bar{C} = O) \cap (B\bar{A} = O) \cap (AC = O) \cap -(B = O) = O,$$

which can also be read

$$(B\bar{C} = O) \cap (B\bar{A} = O) \cap -(B = O) < -(AC = O),$$

or 'if every B is a C, every B is an A, and the class B is not empty, some A is a C.' In this case, then, the conclusion depends on three propositions.

10

Finally, we shall treat several questions relating to logical equations. The identities

$$(A < B) = (B > A) = (A\bar{B} = \bigcirc)$$

and

$$(A = B) = (A < B) \cap (A > B) = (A\bar{B} = \bigcirc) \cap (\bar{A}B = \bigcirc)$$
$$= (A\bar{B} \cup \bar{A}B = \bigcirc)$$

say that each logical equation can be transformed into an equivalent one in which the second member is $\bigcirc$.

The identity

$$(A = \bigcirc) \cap (B = \bigcirc) \cap (C = \bigcirc) \cap \ldots = (A \cup B \cup C \cup \ldots = \bigcirc)$$

says that the system of several coexistent logical equations can be reduced to only one equation, whose second member is the empty set.

An equation or system of equations may contain an unknown class X; we may pose the problem of solving for this unknown. The system of equations having been reduced, for this purpose, to only one whose second member is null, the equation will have the form $f(X) = \bigcirc$, where $f(X)$ represents a logical function of X. By what has been said, $f(X)$ may be put in the separated form $AX \cup B\bar{X}$, and so every equation or system of logical equations can be reduced to the form

$$AX \cup B\bar{X} = \bigcirc$$

or $(AX = \bigcirc) \cap (B\bar{X} = \bigcirc)$,

which can also be written

$$(X < \bar{A}) \cap (B < X)$$

or $B < X < \bar{A}$.

In order for these equations to be possible, we must have $B < \bar{A}$, $AB = \bigcirc$. Now, supposing this condition to be verified, it is sufficient to take for X any class whatever containing B and contained in $\bar{A}$, which can be done, if B and $\bar{A}$ are not equal, in an infinite number of ways. The smallest class X is B and the largest is $\bar{A}$; every other can be put in the form $B \cup Z\bar{A}$, where Z is an arbitrary class.

11

The *elimination* from an equation (or system of equations) of an unknown means writing, where possible, an equation no longer containing the unknown, but containing the other variables in such a way that the proposed equation can be satisfied by any value of the unknown. We have already seen that the result of the elimination of X from the equation $AX \cup B\overline{X} = \bigcirc$ is $AB = \bigcirc$.

The resolution of a system of logical equations with several unknowns can be reduced to the questions already treated by eliminating the unknowns one at a time, as in algebra.

By eliminating a system of variables from one or more equations we get the condition which must hold among the remaining variables in order for the system to be possible. Thus from the equation

$$AXY \cup BX\overline{Y} \cup C\overline{X}Y \cup D\overline{X}\,\overline{Y} = \bigcirc,$$

eliminating X first we have

$$(AY \cup B\overline{Y}) \cap (CY \cup D\overline{Y}) = \bigcirc$$

or $\;\; ACY \cup BD\overline{Y} = \bigcirc;$

eliminating Y from this we have

$$ACBD = \bigcirc$$

as the necessary condition for the proposed equation to be satisfied by certain classes X and Y.

Another application of elimination is the syllogism itself. From two propositions containing three classes (major term, minor term, middle term) we eliminate the middle term in order to have a relation between the other two. Thus, if we want to eliminate B from the propositions $A < B$ and $B < C$, we will have

$$(A < B) \cap (B < C) = (A\overline{B} = \bigcirc) \cap (B\overline{C} = \bigcirc)$$
$$= (A\overline{B} \cup B\overline{C} = \bigcirc).$$

Here the first member is in a separated form, and by eliminating B we have $A\overline{C} = \bigcirc$, which is precisely the conclusion of the syllogism.

On the other hand, from the premises already considered at the end of §9,

$$(B\overline{C} = \bigcirc) \cap (B\overline{A} = \bigcirc),$$

which can be written $B(\overline{C} \cup \overline{A}) = \bigcirc$, and eliminating B we have the identity $\bigcirc = \bigcirc$, thus confirming that from just those two premises no relation between A and C may be concluded.

2 GEOMETRICAL FORMATIONS

Segments, surfaces, and volumes

1

Geometrical calculus consists of a system of operations analogous to those of algebraic calculus, but the entities on which the calculations are carried out, instead of being numbers, are geometrical entities which we shall define.

Let A, B, C, ... be points in space.

DEF. 1 *We shall call the line limited by two points A and B segment AB, and imagine it as described by a point P that moves from A to B.*

Given a segment AB, if A and B do not coincide, a straight line is determined which contains it, and which we shall call line AB. Its length is also determined; by the expression mg AB, which is read *magnitude AB*, we shall understand the number which measures the length of this segment, with respect to a fixed length, which we shall call the unit of measure. Furthermore, given a segment, the direction in which it is imagined to be described is also determined. The segments AB and BA lie on the same line and have the same magnitude, but they have opposite directions. Hence, they are not to be considered as identical.

DEF. 2 *We shall call the triangular planar surface described by the segment AP, where the point P describes the segment BC, from B towards C, surface ABC.*

Given a surface ABC, if A, B, C are not on a line, the plane that contains it, which we shall call plane ABC, is determined. Its area is also determined, and we shall call the number which measures it, with respect to a fixed area assumed as the unit of measure, mg ABC.

Given a surface, the direction in which it is imagined to be described is also determined.

DEF. 3 *We shall call the tetrahedral volume described by the surface ABP, where the point P describes the segment CD, in the direction already agreed on, volume ABCD.*

By mg $ABCD$ we understand the number which measures this volume, with respect to a fixed volume assumed as the unit of measure.

2

Often we shall represent a segment, or a surface, or a volume, by a single letter. We shall most often use the letters a, b, ... for segments; α, β, ... for surfaces; and A, B, Γ, ... for volumes.

If a, a', ... represent the segments PG, $P'G'$, ... and if α represents the surface XYZ, to the expressions

Aa, aA; ABa, AaB, aAB, αA, $A\alpha$, aa'

we shall attribute the meanings, respectively,

APQ, PQA; $ABPQ$, $APQB$, $PQAB$, $X\overset{\cdot}{Y}ZA$, $AXYZ$, $PQP'Q'$.

Operations on volumes

3

DEF. 1 *We shall say that a volume* A *is* zero, *and write* A $= 0$, *if its magnitude is zero.*

This convention already allows us to express several geometrical propositions in a more concise form. Thus:

'$ABCD = 0$' means 'the points $ABCD$ lie in the same plane';

'$A\alpha = 0$' means 'the point A lies in plane α';

'$ab = 0$' means 'the lines a and b lie in the same plane, i.e., either they meet or are parallel.'

DEF. 2 *A volume* $ABCD$, *which is non-zero, is said to be* right-handed, *or* described in the positive direction, *if a person situated on* AB, *with head at* A *and lower extremities at* B, *sees the surface* ABP *that describes the volume when the point* P *moves along the segment* CD, *from* C *towards* D, *from left to right. If, instead, the same person sees the plane as moving from right to left, the volume is said to be* left-handed, *or* described in the negative direction.

Thus, for example, the volume $ABCD$, where B, C, D have the positions shown in the figure, will be right-handed if A is in front of the plane of the page, and left-handed otherwise. The direction of a volume, which is rarely considered in elementary geometry, has great importance in the research we are undertaking.

DEF. 3 *We shall designate as* ratio of two volumes A and Ω, *indicated by* A$/\Omega$, *the ratio of their magnitudes taken with the* $+$ *sign or with the* $-$ *sign, according as the two volumes have the same direction or opposite directions.*

Hence, the ratio of two volumes is a real number, which may be positive or negative.

Let Ω be a volume fixed in magnitude and direction. Then:

DEF. 4 *We shall say that the volumes* A *and* B *are equal, and write* A $=$ B, *if*

A$/\Omega = $ B$/\Omega$.

DEF. 5 *We shall say that the volume* A *is equal to the volume* B *multiplied by the number m, and write* A $= m$B, *or* A $= $ Bm, *if*

A$/\Omega = m($B$/\Omega)$.

DEF. 6 *We shall say that the volume* A *is the sum of volumes* B *and* Γ, *and write* A = B + Γ, *if*

$$\frac{A}{\Omega} = \frac{B}{\Omega} + \frac{\Gamma}{\Omega}.$$

4

By virtue of these definitions a well-defined meaning has been given to every equality of the form

$$\Sigma = m\text{A} + n\text{B} + p\Gamma + ...,$$

in which m, n, p, ... are numbers and A, B, Γ, ... are volumes. It is equivalent to the numerical equality

$$\frac{\Sigma}{\Omega} = m\frac{A}{\Omega} + n\frac{B}{\Omega} + p\frac{\Gamma}{\Omega} + ...$$

Hence, the same identities hold for volumes as for numbers, and so we have

$$A + B = B + A, \qquad A + (B + \Gamma) = A + B + \Gamma,$$
$$(m + n)A = mA + nA, \qquad m(A + B) = mA + mB,$$
$$(A = B) < (A + \Gamma = B + \Gamma), \qquad (A = B) < (mA = mB),$$
$$(A = B) \cap (B = \Gamma) < (A = \Gamma),$$

which express the commutative and associative properties of addition for volumes, the distributive property of the product of a number by a volume, with respect to both factors, and the uniformity of these operations.

We must carefully distinguish the equalities mg A = mg B and A = B; the first says that two volumes have the same magnitude, and the second that they have the same magnitude and the same direction.

In conformity with algebraic notation, to the expressions −A and A − B we shall attribute the meanings (−1)A, A + (−1)B. Hence the equality A = −B says that the volumes A and B have the same magnitude but opposite direction.

The inequality $\pi A/\pi B > 0$ means that the volumes πA and πB have the same direction, or that the points A and B lie on the same side of the plane π. Analogously,

$(\pi A/\pi B < 0) =$ (the points A and B lie on opposite sides of the plane π),

$(\pi A = \pi B) =$ (the line AB is parallel to the plane π).

Indeed, if $\pi A = \pi B$, the tetrahedrons πA and πB, which have the same base, and equal magnitudes, will have the same height; and since they have the same direction, the points A and B lie on the same side of the plane π, and hence the line AB is parallel to π; and vice versa.

If in a volume $ABCD$ we interchange two vertices, the magnitude of the volume does not change, but, evidently, the direction changes.

Interchanging one or more vertices, the following identities may be deduced:

$$ABCD = -BACD, \qquad ABCD = BCAD,$$
$$ABCD = -BCDA, \qquad ABCD = CDAB,$$

which may also be written, on changing letters,

$$ABa = -BAa, \qquad AaD = aAD, \qquad A\alpha = -\alpha A, \qquad ab = ba.$$

Geometrical formations

5

DEF. 1 *The set of points A, B, C, ... to which are affixed, respectively, the numbers m, n, p, ... is said to be a* formation of the first species, *and is represented by the expression $mA + nB + pC + ...$*

DEF. 2 *The set of segments $a, b, c, ...$ to which are affixed the numbers $m, n, p, ...$ is said to be a* formation of the second species, *and is represented by the expression $ma + nb + pc + ...$*

DEF. 3 *The set of surfaces α, β, γ, ... to which are affixed the numbers $m, n, p, ...$ is said to be a* formation of the third species, *and is represented by the expression $m\alpha + n\beta + p\gamma + ...$*

DEF. 4 *Volumes and sums of volumes of the form $mA + nB + p\Gamma + ...$ are said to be* formations of the fourth species.

The formations of the four species, just defined, are the geometrical entities which form the object of our calculus. For those of the fourth species several operations have already been defined; we shall make the analogous definitions for the first three species depend on those of the fourth in the following way:

6

DEF. *A formation of the* $\begin{Bmatrix} 1st \\ 2nd \\ 3rd \end{Bmatrix}$ *species* $\begin{Bmatrix} mA + nB + ... \\ ma + nb + ... \\ m\alpha + n\beta + ... \end{Bmatrix}$ *is said to be zero,*

$$and\ we\ write\ \begin{Bmatrix} mA + nB + ... = 0 \\ ma + nb + ... = 0 \\ m\alpha + n\beta + ... = 0 \end{Bmatrix},\ if,\ however\ we\ take\ the\ \begin{Bmatrix} surface\ \pi \\ segment\ p \\ point\ P \end{Bmatrix},$$

$$we\ always\ have\ \begin{Bmatrix} mA\pi + nB\pi + ... = 0 \\ map + nbp + ... = 0 \\ m\alpha P + n\beta P + ... = 0 \end{Bmatrix}.$$

Hence, the equality $ABC = 0$ means that the points ABC have such a position that however we take the point P we have $ABCP = 0$, which evidently means that the area of the triangle ABC is zero. Hence

$(ABC = 0) = $ (the points A, B, C lie on the same line),

$(Aa = 0) = $ (the point A lies on the line a).

The equality $AB = 0$ means that, however we take the points P and Q, we have $ABPQ = 0$, which evidently means that the points A and B coincide. Hence

$(AB = 0) = $ (the points A and B coincide).

7

DEF. *Two formations of the* $\begin{Bmatrix} 1st \\ 2nd \\ 3rd \end{Bmatrix}$ *species* $\begin{Bmatrix} mA + nB + ... \\ ma + nb + ... \\ m\alpha + n\beta + ... \end{Bmatrix}$ *and*

$\begin{Bmatrix} m'A' + n'B' + ... \\ m'a' + n'b' + ... \\ m'\alpha' + n'\beta' + ... \end{Bmatrix}$ *are said to be* equal, *and we write*

$$\begin{Bmatrix} mA + nB + ... = m'A' + n'B' + ... \\ ma + nb + ... = m'a' + n'b' + ... \\ m\alpha + n\beta + ... = m'\alpha' + n'\beta' + ... \end{Bmatrix},\ if,\ however\ we\ take\ the$$

$\begin{Bmatrix} surface\ \pi \\ segment\ p \\ point\ P \end{Bmatrix}$, *we always have* $\begin{Bmatrix} mA\pi + nA\pi + ... = m'B'\pi + n'B'\pi + ... \\ map + nbp + ... = m'a'p + n'b'p + ... \\ m\alpha P + n\beta P + ... = m'\alpha'P + n'\beta'P + ... \end{Bmatrix}$.

From the preceding definition we deduce:

THEOREM 1 *Two formations of the same species, which differ only in the order of their terms, are equal to each other.*

In fact, letting each term of each formation be followed by a surface π, or by a segment p, or by a point P, according to whether it is of the first, second, or third species, we have two sums of volumes, which differ only in the order of their terms, and hence are equal. Therefore, the proposed formations are also equal.

THEOREM 2 *Two formations equal to a third are equal to each other.*

The proof is analogous to the preceding.

8

The reader may easily prove the following propositions.

THEOREM 1 *The equation* $mA = nB$ *says that the points A and B coincide, and that the numbers m and n are equal.*

DEF. *Two segments a and b which lie on the same line are said to have the same direction, or opposite directions, according to whether for two points P and Q, arbitrarily chosen, the volumes aPQ and bPQ have the same direction or opposite directions.*

THEOREM 2 *The equation* $a = b$ *says that the segments a and b lie on the same line, have equal magnitudes, and have the same direction.*

THEOREM 3 *The equation* $a = mb$ *says that the segments a and b lie on the same line, that the ratio of their magnitudes is equal to the absolute value of m, and that they have the same direction, or opposite directions, according as m is positive or negative.*

DEF. *Two surfaces* α *and* β *which lie in the same plane are said to have the same direction, or opposite directions, according to whether for an arbitrary point P the volumes* αP *and* βP *have the same or opposite directions.*

THEOREM 4 *The equation* $\alpha = \beta$ *says that the surfaces* α *and* β *lie in the same plane, have equal magnitudes, and have the same direction.*

COROLLARY *The equation* $ABC = ABD$ *says that the line CD is parallel to AB.*

THEOREM 5 *The equation* $\alpha = m\beta$ *says that the surfaces* α *and* β *lie in the same plane, that the ratio of their magnitudes is equal to the absolute value of m, and that they have the same direction, or opposite directions, according as m is positive or negative.*

We have the identities

$$AB = -BA, \qquad Aa = +aA.$$

To prove them it is sufficient to follow the members of the first with p, and those of the second with P, thus obtaining the known identities

$$ABp = -BAp, \qquad AaP = aAP.$$

Operations on formations

9

We now introduce the following operations on formations. For simplicity, by the letters A, B, C, ... we understand indifferently points, or segments, or surfaces, or volumes; by the letters S, S', ... we understand formations of any species whatever.

DEF. 1 *We shall designate as* sum *of two formations of the same species* $S = mA + nB + ...$ *and* $S' = m'A' + n'B' + ...$, *and indicate by* $S + S'$, *the formation*

$$mA + nB + ... + m'A' + n'B' + ...$$

composed of the terms of the two given formations.

This gives the identities

$$S + S' = S' + S, \qquad S + (S' + S'') = S + S' + S''.$$

Indeed, the two members of these identities contain the same terms.

DEF. 2 *We shall designate as* product *of a number x and a formation* $S = mA + nB + ...$, *and indicate by xS, the formation* $xmA + xnB + ...$ *which we obtain by multiplying the coefficient of each of its terms by x.*

We have the identities

$$x(S + S') = xS + xS', \qquad (x + y)S = xS + yS$$

because the members of these identities contain the same terms.

It is quite easy to prove the following propositions:

$$(S = S') < (S + S'' = S' + S''),$$
$$(S = S') \cap (S'' = S''') < (S + S'' = S' + S'''),$$
$$(S = S') < (xS = xS'),$$

which say that geometrical equations may be summed and multiplied by numbers, just like algebraic equations.

We shall continue to make use of the notations represented by the following equations, in which S, S' are formations and x is a number:

$$Sx = xS,$$
$$-S = (-1)S,$$
$$S - S' = S + (-S'),$$
$$(S/S' = x) = (S = xS').$$

The last, which is a logical equation, may be stated: we shall designate as ratio of two formations S and S' that number x which, when multiplied by S', gives S. In order for the ratio S/S' to have meaning it is first of all necessary for the two formations to be of the same species; but this condition is not sufficient. It results from what has been said that the following have ratios: two volumes, two surfaces which lie in the same plane, and two segments which lie on the same line.

10

The preceding operations are the natural extension of the analogous algebraic operations. We shall now define a geometrical operation which has no correspondent in algebra.

DEF. *Let* $S = m_1 A_1 + m_2 A_2 + ... + m_n A_n$ *and* $S' = m_1' A_1' + ... + m_n' A_n'$ *be two formations of species s and s', and suppose that* $s + s' \leqslant 4$. *Then we shall designate as* formation projecting from S to S', *or* progressive product *of the two formations, and indicate by* SS', *the formation* $\sum_{i,j} m_i m_j' A_i A_j'$, *i.e. the sum of all the terms obtained from the expression by substituting for i and j all whole numbers from 1 to n and from 1 to n', respectively.*

We have, for example:

$$(A + B)C = AB + AC, \qquad (B - A)\pi = B\pi - A\pi,$$
$$((B - A)\pi = 0) = (\text{the line } AB \text{ is parallel to the plane } \pi),$$
$$((B - A)a = 0) = (\text{the line } AB \text{ is parallel to the line } a).$$

The definition of §6 can now be stated: A formation of the 1st, or 2nd, or 3rd species is said to be zero if, projecting from it, respectively, a surface, a segment, or a point, arbitrarily chosen, the sum of the volumes so obtained is zero. The definition of §7 may also be stated analogously.

We have the following identities:

$$S(S_1 + S_2) = SS_1 + SS_2, \qquad (S_1 + S_2)S = S_1 S + S_2 S,$$
$$S(S_1 S_2) = SS_1 S_2,$$
$$(xS)S' = S(xS') = x(SS').$$

Indeed the two members of each of these equalities contain the same terms, at most in a different order.

If the formations S and S' are of species s and s', we have

$$SS' = (-1)^{ss'} S'S.$$

Indeed the two formations SS' and $S'S$ contain the same terms, with the same sign if ss' is even and with opposite signs if ss' is odd.

We now have the following theorem.

THEOREM *If we multiply member by member two geometrical equations, we obtain a new equation which is a consequence of those two; i.e.,*

$$(S = S') \cap (S_1 = S_1') < (SS_1 = S'S_1').$$

To make the proof easier, we shall suppose that all formations are of the first species and begin with several particular cases.

(a) $(S = S') < (SP = S'P)$.

In fact, $(S = S') =$ (whatever the surface π we have $S\pi = S'\pi$) < (whatever the segment p we have $SPp = S'Pp$) = $(SP = S'P)$, which proves (a).

(b) $(S = S') < (SS_1 = S'S_1)$.

In fact, letting $S_1 = m_1A_1 + n_1B_1 + ...$, we have by (a)

$$(S = S') < (SA_1 = S'A_1) < (m_1SA_1 = m_1S'A_1),$$
$$(S = S') < \ . \ . \ . \ . \ . \ . \ . \ . \ < (n_1SB_1 = n_1S'B_1),$$

$$. \ .$$

Multiplying these logical equations, we have

$$(S = S') < (m_1SA_1 = m_1S'A_1) \cap (m_2SA_2 = m_2S'A_2) \cap ... <$$
$$(m_1SA_1 + m_2SA_2 + ... = m_1S'A_1 + m_2S'A_2 + ...) = (SS_1 = S'S_1),$$

which proves (b).

(c) $(S_1 = S_1') < (SS_1 = SS_1')$.

In fact

$$(S_1 = S_1') < (S_1S = S_1'S) = (-SS_1 = -SS_1') = (SS_1 = SS_1').$$

(d) $(S = S') \cap (S_1 = S_1') < (SS_1 = S'S_1')$.

In fact

$$(S = S') < (SS_1 = S'S_1),$$
$$(S_1 = S_1') < (S'S_1 = S'S_1')$$

and hence

$$(S = S') \cap (S_1 = S_1') < (SS_1 = S'S_1) \cap (S'S_1 = S'S_1')$$
$$< (SS_1 = S'S_1').$$

11

To sum up what we have said, we have introduced four species of formations, of which particular cases are points, segments, surfaces, and volumes.

We have defined (§9) the sum of two formations of the same species, which operation has commutative and associative properties like algebraic addition.

We have defined the product of a number by a formation of any species whatever, and this operation is distributive with respect to both factors.

Finally, we have defined the projection, or progressive product, of two formations of species s and s', assuming $s + s' \leqslant 4$, which is a new formation of species $s + s'$. This has the distributive property with respect to both factors and is associative; on interchanging the factors the product becomes multiplied by $(-1)^{ss'}$. The product of a number by a formation could be considered as a product of two formations by considering numbers as formations of zero species.

The equality of two formations of the fourth species was made to depend on the equality of two numbers. It was then assumed by definition that two formations of the first, second, or third species are equal if, by following them with the same surface, or the same segment, or the same point, the volumes obtained are equal. Finally it was proved that if two formations are equal, then preceding them or following them by the same formation, or by equal formations, yields equal results.

We shall next consider the reduction of formations of successive species to their simplest form, using, for this purpose, a few propositions of geometry and the geometrical calculus just explained.

VII

The principles of arithmetic,
presented by a new method (1889)*

This remarkable little booklet (given here in its entirety) contains the first statement of Peano's best-known achievement, the postulates for the natural numbers. The essential ideas were first published by Dedekind, as Peano very generously points out in a later work, but there can be no doubt of the originality of Peano's work, and we have his own statement that he saw Dedekind's work only when Arithmetices principia *was going to the press.*

Following the preface, there is an introductory section on logical notation. In it we find, for the first time, the symbol ε to indicate membership and ⊃ to indicate inclusion, and a note on the necessity to distinguish these two. The first section of the Arithmetices principia *proper contains the famous five postulates, along with four others dealing with the sign* =. *These postulates were slightly modified in later writings.*

The bulk of the work is written in mathematical and logical symbols. The preface and explanatory notes are in Latin, and the title is Latin, or almost – 'arithmetices' is transliterated Greek (the completely Latinized form would be 'arithmeticae').

PREFACE

Questions pertaining to the foundations of mathematics, although treated by many these days, still lack a satisfactory solution. The difficulty arises principally from the ambiguity of ordinary language. For this reason it is of the greatest concern to consider attentively the words we use. I resolved to do this, and am presenting in this paper the results of my study with applications to arithmetic.

I have indicated by signs all the ideas which occur in the fundamentals of arithmetic, so that every proposition is stated with just these signs. The signs pertain either to logic or to arithmetic. The signs of logic that occur here are about ten in number, although not all are necessary. The use of these signs and several of their properties are explained in ordinary language

* *Arithmetices principia nova methodo exposita* (Turin: Bocca, 1889) [16].

in the first part. I did not wish to present their theory more fully here. The signs of arithmetic are explained as they occur.

With this notation every proposition assumes the form and precision equations enjoy in algebra, and from propositions so written others may be deduced, by a process which resembles the solution of algebraic equations. That is the chief reason for writing this paper.

Having made up the signs with which I can write arithmetical propositions, in treating them I have used a method which, because it is to be followed in later studies, I shall present briefly.

Those arithmetical signs which may be expressed by using others along with signs of logic represent the ideas we can define. Thus I have defined every sign, if you except the four which are contained in the explanations of §1. If, as I believe, these cannot be reduced further, then the ideas expressed by them may not be defined by ideas already supposed to be known.

Propositions which are deduced from others by the operations of logic are *theorems*; those for which this is not true I have called *axioms*. There are nine axioms here (§1), and they express fundamental properties of the undefined signs.

In §§1–6 I have proved the ordinary properties of numbers; for the sake of brevity, I have omitted proofs which are similar to preceding ones. The ordinary form of proofs has had to be altered in order that they may be expressed with the signs of logic. This transformation is sometimes rather difficult but the nature of the proof then becomes quite clear.

In the following sections I have treated various things so that the power of the method is better seen. In §7 are several theorems pertaining to the theory of numbers. In §§8 and 9 are found the definitions of rationals and irrationals. Finally, in §10 I have given several theorems, which I believe to be new, pertaining to the theory of those entities which Professor Cantor has called *Punktmenge* (*ensemble de points*).

In this paper I have used the research of others. The notations and propositions of logic which are contained in numbers II, III, and IV, with some exceptions, represent the work of many, among them Boole especially.[1] The sign ε, which must not be confused with the sign ⊃, applications of the

---

1 Boole, *The Mathematical Analysis of Logic* ... (Cambridge, 1847); 'The calculus of logic,' *Camb. and Dublin Math. J.*, 3 (1848), 183–98; *An Investigation of the Laws of Thought* etc. (London, 1854). E. Schröder, *Der Operationskreis des Logikkalkuls* (Leipzig, 1877) – he had already treated several matters pertaining to logic in a preceding work; *Lehrbuch der Arithmetik und Algebra*, etc. (Leipzig, 1873). I gave a very brief presentation of the theories of Boole and Schröder in my book *Calcolo geometrico* etc. (Torino, 1888). See: C.S. Peirce, 'On the algebra of logic,' *Amer. J. Math.*, 3 (1880), 15–57; 7 (1885), 180–202; Jevons, *The Principles of Science* (London, 1883); MacColl, 'The calculus of equivalent statements,' *Proc. London Math. Soc.* 9 (1877), 9–20; 10 (1878), 16–28.

inverse in logic, and a few other conventions, I have adopted so that I could express any proposition whatever. In the proofs of arithmetic I used the book H. Grassmann, *Lehrbuch der Arithmetik* (Berlin, 1861). Also quite useful to me was the recent work by R. Dedekind, *Was sind und was sollen die Zahlen* (Braunschweig, 1888), in which questions pertaining to the foundations of numbers are acutely examined.

My booklet should be taken as a sample of this new method. With these notations we can state and prove innumerable other propositions, such as those which pertain to rationals and irrationals. But in order to treat other theories, it is necessary to adopt new signs to indicate new entities. I believe, however, that with only these signs of logic the propositions of any science can be expressed, so long as the signs which represent the entities of the science are added.

TABLE OF SIGNS

Signs of logic

Sign	Meaning	Page [of the original edition]
P	proposition	VII
K	class	X
∩	and	VII, X
∪	or	VIII, X, XI
−	not	VIII, X
Λ	false *or* nothing	VIII, XI
Ɔ	one deduces *or* is contained in	VIII, XI
=	equals	VIII
ε	is	X
[ ]	*sign of the inverse*	XI
Ɛ	such that *or* [ε]	XII
Th	Theorem	XVI
Hp	Hypothesis	XVI
Ts	Thesis	XVI
L	Logic	XVI

Signs of arithmetic

The signs 1, 2, ..., =, >, <, +, −, × have their usual meaning. The sign of division is /.

Sign	Meaning	Page [of the original edition]
N	positive integers	1
R	positive rational numbers	12

Q	quantity or positive real numbers	16
Np	prime number	9
M	maximum	6
И	minimum	6
T	terminus or greatest bound	15
D	divides	9
ɑ	is divisible	9
π	is prime with	9

Composite signs

− <	is not less than
= ∪ >	is equal to or greater than
₃D	is a divisor
M₃D	is the greatest divisor

NOTATIONS OF LOGIC

I *Punctuation*

By the letters a, b, ..., x, y, ..., x', y', ... we indicate any indeterminate entities. Determinate entities are, however, indicated by the signs, or rather by the letters, P, K, N, ...

Generally we write signs on the same line. So that it will be clear how they are to be joined, we use parentheses, as in algebra, or rather points . : ∴ :: etc. So that a formula divided by points may be understood, first the signs which are not separated by points are taken together, then those separated by one point, then those by two points, etc. For example, let a, b, c, ... be any signs. Then $ab . cd$ means $(ab)(cd)$; and $ab . cd : ef . gh ∴ k$ means $(((ab)(cd))((ef)(gh)))k$.

The signs of punctuation may be omitted if formulas having different punctuation have the same meaning, or if just one formula, that being the one we wish to write, has meaning. To avoid the danger of ambiguity, we never use . , : as signs of arithmetical operations.

II *Propositions*

The sign P means *proposition*.

The sign ∩ is read *and*. Let a, b be propositions; then $a ∩ b$ is the simultaneous affirmation of the propositions a, b. For the sake of brevity, instead of $a ∩ b$, we ordinarily write ab.

The sign $-$ is read *not*. Let a be a P; then $-a$ is the negation of the proposition a.

The sign $\cup$ is read *or*. Let a, b be propositions; then $a \cup b$ is the same as $- : -a \, . -b$.

(The sign V means *true*, or *identity*, but we never use this sign.)

The sign $\wedge$ means *false*, or *absurd*.

(The sign C means *is a consequence of*. Thus b C a is read b *is a consequence of the proposition a*. But we never use this sign.)

The sign $\supset$ means *one deduces*. Thus $a \supset b$ means the same as b C a. If the propositions a, b contain the indeterminate quantities x, y, ..., that is, express conditions on these objects, then $a \supset_{x,y,...} b$ means: whatever the x, y, ..., from proposition a one deduces b. If indeed there is no danger of ambiguity, instead of $\supset_{x,y,...}$ we write only $\supset$.

III *Propositions of logic*

Let a, b, c, ... be propositions. We have:

1. $a \supset a$.
2. $a \supset b . b \supset c : \supset : a \supset c$.
3. $a = b . = : a \supset b . b \supset a$.
4. $a = a$.
5. $a = b . = . b = a$.
6. $a = b . b \supset c : \supset . a \supset c$.
7. $a \supset b . b = c : \supset . a \supset c$.
8. $a = b . b = c : \supset . a = c$.
9. $a = b . \supset . a \supset b$.
10. $a = b . \supset . b \supset a$.

11. $ab \supset a$.
12. $ab = ba$.
13. $a(bc) = (ab)c = abc$.
14. $aa = a$.
15. $a = b . \supset . ac = bc$.
16. $a \supset b . \supset . ac \supset bc$.
17. $a \supset b . c \supset d : \supset . ac \supset bd$.
18. $a \supset b . a \supset c : = . a \supset bc$.
19. $a = b . c = d : \supset . ac = bd$.
20. $-(-a) = a$.

21. $a = b \, . \, = \, . \, -a = -b.$

22. $a \supset b \, . \, = \, . \, -b \supset -a.$

23. $a \cup b \, . \, = \, \therefore \, - \, : \, - a \, . \, - b.$

24. $-(ab) = (-a) \cup (-b).$

25. $-(a \cup b) = (-a)(-b).$

26. $a \supset . \, a \cup b.$

27. $a \cup b = b \cup a.$

28. $a \cup (b \cup c) = (a \cup b) \cup c = a \cup b \cup c.$

29. $a \cup a = a.$

30. $a(b \cup c) = ab \cup ac.$

31. $a = b \, . \supset . \, a \cup c = b \cup c.$

32. $a \supset b \, . \supset . \, a \cup c \supset b \cup c.$

33. $a \supset b \, . \, c \supset d : \supset : a \cup c \, . \supset . \, b \cup d.$

34. $b \supset a \, . \, c \supset a : = . \, b \cup c \supset a.$

35. $a - a = \Lambda.$

36. $a \Lambda = \Lambda.$

37. $a \cup \Lambda = a.$

38. $a \supset \Lambda \, . \, = . \, a = \Lambda.$

39. $a \supset b \, . \, = . \, a - b = \Lambda.$

40. $\Lambda \supset a.$

41. $a \cup b = \Lambda \, . \, = \, : a = \Lambda \, . \, b = \Lambda.$

42. $a \supset . \, b \supset c : = \, : ab \supset c.$

43. $a \supset . \, b = c \, : = . \, ab = ac.$

Let α be the sign of some relation (e.g., $=, \supset$) so that $a \, \alpha \, b$ is a proposition. Then in place of $- . \, a \, \alpha \, b$, we write $a - \alpha \, b$. Thus:

$$a - = b \, . \, = \, : - . \, a = b,$$
$$a - \supset b \, . \, = \, : - . \, a \supset b.$$

Thus the sign $- =$ means *is not equal to*. If the proposition a contains the indeterminate x, $a \, - =_x \Lambda$ means: there is an x which satisfies condition a. The sign $- \supset$ means *one does not deduce*.

Similarly, if α and β are signs of relations, in place of $a \, \alpha \, b \, . \, a \, \beta \, b$, and $a \, \alpha \, b \, . \cup . \, a \, \beta \, b$ we may write $a \, . \, \alpha \beta \, . \, b$ and $a \, . \, \alpha \cup \beta \, . \, b$. Thus, if a and b

are propositions, the formula $a . \supset - = . b$ says: from a one deduces b, but not vice versa:

$$a . \supset - = . b : = : a \supset b . b - \supset a.$$

Formulas:

$$a \supset b . b \supset c . a - \supset c : = \Lambda.$$
$$a = b . b = c . a - = c : = \Lambda.$$
$$a \supset b . b \supset - = c : \supset . a \supset - = c.$$
$$a \supset - = b . b \supset c : \supset . a \supset - = c.$$

But we shall rarely use these notations.

IV *Classes*

The sign K means *class*, or aggregate of entities.

The sign ε means *is*. Thus $a \, \varepsilon \, b$ is read *a is a b*; $a \, \varepsilon \,$ K means *a is a class*; $a \, \varepsilon \,$ P means *a is a proposition*.

In place of $-(a \, \varepsilon \, b)$ we shall write $a - \varepsilon \, b$. The sign $-\varepsilon$ means *is not*. Thus:

44. $a - \varepsilon \, b . = : - . a \, \varepsilon \, b.$

The sign $a, b, c \, \varepsilon \, m$ means: a, b, and c are in m. Thus:

45. $a, b, c \, \varepsilon \, m . = : a \, \varepsilon \, m . b \, \varepsilon \, m . c \, \varepsilon \, m.$

Let a be a class. Then $-a$ means that class made up of individuals that are not in a.

46. $a \, \varepsilon \,$ K $. \supset : x \, \varepsilon \, -a . =_x . x - \varepsilon \, a.$

Let a, b be classes. Then $a \cap b$, or ab, is the class composed of individuals which are at the same time in a and b; $a \cup b$ is the class composed of individuals which are in a or b.

47. $a, b \, \varepsilon \,$ K $. \supset \therefore x \, \varepsilon . ab : =_x : x \, \varepsilon \, a . x \, \varepsilon \, b.$

48. $a, b \, \varepsilon \,$ K $. \supset \therefore x \, \varepsilon . a \cup b : =_x : x \, \varepsilon \, a . \cup . x \, \varepsilon \, b.$

The sign Λ indicates the class which contains no individuals. Thus:

49. $a \, \varepsilon \,$ K $. \supset \therefore a = \Lambda : = : x \, \varepsilon \, a . =_x \Lambda.$

(We shall not use the sign V, which indicates the class composed of all individuals being considered.)

The sign $\supset$ means *is contained*. Thus $a \supset b$ means *the class a is contained in the class b*.

50. $a, b \, \varepsilon \, K . \supset \, \therefore \, a \supset b : = \, : x \, \varepsilon \, a . \supset_x . x \, \varepsilon \, b.$

(The formula $b \, C \, a$ could mean *the class b contains the class a*, but we shall not use the sign C.)

The signs Λ and $\supset$ have meanings here which are slightly different from the preceding, but no ambiguity will arise, for if propositions are being considered, the signs are read *absurd* and *one deduces*, but if classes are being considered, they are read *empty* and *is contained*.

The formula $a = b$, if a and b are classes, means $a \supset b . b \supset a$. Thus

51. $a, b \, \varepsilon \, K . \supset \, \therefore \, a = b : = \, : x \, \varepsilon \, a . =_x . x \, \varepsilon \, b.$

Propositions 1, ..., 41 also hold if $a, b, ...$ indicate classes. In addition, we have:

52. $a \, \varepsilon \, b . \supset . b \, \varepsilon \, K.$
53. $a \, \varepsilon \, b . \supset . b - = \Lambda.$
54. $a \, \varepsilon \, b . b = c : \supset . a \, \varepsilon \, c.$
55. $a \, \varepsilon \, b . b \supset c : \supset . a \, \varepsilon \, c.$

Let s be a class, and k be a class which is contained in s; then we say that k is an individual of the class s if k consists of only one individual. That is:

56. $s \, \varepsilon \, K . k \supset s : \supset \, :: k \, \varepsilon \, s . = \, \therefore \, k - = \Lambda : x, y \, \varepsilon \, k . \supset_{x,y} . x = y.$

v *The inverse*

The sign of the inverse is [ ], and we shall explain its use in the following section. Here we give some particular examples.

1. Let a be a proposition containing the indeterminate x; then the expression $[x \varepsilon]a$, which is read *those x such that a*, or *solutions*, or *roots* of the condition a, indicates the class consisting of individuals which satisfy the condition a. That is:

57. $a \, \varepsilon \, P . \supset : [x \varepsilon]a . \varepsilon \, K.$
58. $a \, \varepsilon \, K . \supset \, \therefore \, [x \varepsilon] . x \, \varepsilon \, a : = a.$
59. $a \, \varepsilon \, P . \supset \, \therefore \, x \, \varepsilon . [x \varepsilon]a : = a.$

Let α, β be propositions containing the indeterminate x. We will have:

60. $[x \varepsilon](\alpha\beta) = ([x \varepsilon]\alpha)([x \varepsilon]\beta).$
61. $[x \varepsilon] - \alpha = -[x \varepsilon]\alpha.$

62. $[x\varepsilon](\alpha \cup \beta) = [x\varepsilon]\alpha \cup [x\varepsilon]\beta.$

63. $\alpha \supset_x \beta . = . [x\varepsilon]\alpha \supset [x\varepsilon]\beta.$

64. $\alpha =_x \beta . = . [x\varepsilon]\alpha = [x\varepsilon]\beta.$

2. Let x, y be any entities. We shall consider the system composed of the entity x and the entity y as a new entity, and indicate it by the sign (x, y); and similarly if the number of entities becomes larger. Let α be a proposition containing the indeterminates x, y; then $[(x, y)\varepsilon]\alpha$ indicates the class of entities (x, y) which satisfy the condition α. We have:

65. $\alpha \supset_{x,y} \beta . = . [(x, y)\varepsilon]\alpha \supset [(x, y)\varepsilon]\beta.$

66. $[(x, y)\varepsilon]\alpha - = \Lambda . = \therefore [x\varepsilon] . [y\varepsilon]\alpha - = \Lambda : - = \Lambda.$

3. Let $x \alpha y$ be a relation between the indeterminates x and y (e.g., in logic, the relations $x = y, x - = y, x \supset y$; in arithmetic $x < y, x > y$, etc.). Then we indicate by the sign $[\varepsilon]\alpha y$ those x which satisfy the relation $x \alpha y$. For convenience, in place of $[\varepsilon]$, we use the sign э. Thus $э \alpha y . = : [x\varepsilon]$. $x \alpha y$, and the sign э is read *the entities such that*. For example, let y be a number; then $э < y$ indicates the class composed of the numbers x which satisfy the condition $x < y$, that is, which are *less than y*. Similarly, since the sign D means *divides*, or *is a divisor*, the formula $э$D means *those which divide* or *the divisors of*. We deduce that $x \varepsilon \, э \alpha y = x \alpha y$.

4. Let α be a formula containing the indeterminate x. Then the expression $x'[x]\alpha$, which is read x' *substituted in place of x in* α, indicates the formula which is obtained if in α, in place of x, we read x'. We deduce that $x[x]\alpha = \alpha$.

5. Let α be a formula which contains the indeterminates $x, y, \ldots$ Then

$$(x', y', \ldots)[x, y, \ldots]\alpha,$$

which is read $x', y', \ldots$ substituted in place of $x, y, \ldots$ in α, indicates the formula which is obtained if in α, in place of $x, y, \ldots$, the letters $x', y', \ldots$ are written. We deduce that $(x, y)[x, y]\alpha = \alpha$.

VI *Functions*

The preceding notations of logic suffice for the expression of any arithmetical proposition, and we shall use only these. Here we briefly explain several other notations which could be useful.

Let s be a class. We suppose an equality has been defined for the entities of the system s which satisfies the conditions:

$a = a,$

$a = b . = . b = a,$

$a = b . b = c : \supset . a = c.$

Let φ be a sign, or aggregate of signs, such that if x is an entity of the class s, then the expression φx indicates a new entity, and suppose an equality has also been defined for the entities φx. Now, if x and y are entities of the class s, and if $x = y$, suppose we can deduce φx = φy. Then the sign φ is said to be a *presign of a function on the class s*, and we write φ ε F's.

$$s \, \varepsilon \, \text{K} . \supset :: \varphi \, \varepsilon \, \text{F}'s . = \, \therefore \, x, y \, \varepsilon \, s . x = y : \supset_{x,y} . \varphi x = \varphi y.$$

If, on the other hand, x is an entity of the class s, the expression xφ indicates a new entity, and from $x = y$ is deduced xφ = yφ, then we say that φ is a *postsign of a function on the class s*, and we write φ ε s'F.

$$s \, \varepsilon \, \text{K} . \supset :: \varphi \, \varepsilon \, s'\text{F} . = \, \therefore \, x, y \, \varepsilon \, s . x = y : \supset_{x,y} . x\varphi = y\varphi.$$

EXAMPLE Let a be a number; then $a+$ is a functional presign on the class of numbers, and $+a$ is a functional postsign, for, whatever the number x, the formulas $a + x$ and $x + a$ indicates new numbers, and from $x = y$ we deduce that $a + x = a + y$ and $x + a = y + a$. Thus

$$a \, \varepsilon \, \text{N} . \supset : a + . \varepsilon . \text{F}'\text{N},$$
$$a \, \varepsilon \, \text{N} . \supset : + a . \varepsilon . \text{N}'\text{F}.$$

Let φ be a functional presign on the class s. Then [φ]y indicates the class composed of the x which satisfy the condition φ$x = y$. That is:

DEF. $s \, \varepsilon \, \text{K} . \varphi \, \varepsilon \, \text{F}'s : \supset : [\varphi]y . = . [x\varepsilon](\varphi x = y).$

The class [φ]y may contain one or more, or even no individuals. We have

$$s \, \varepsilon \, \text{K} . \varphi \, \varepsilon \, \text{F}'s : \supset : y = \varphi x . = . x \varepsilon [\varphi]y.$$

But if φy consists of only one individual, we have $y = \varphi x . = . x = [\varphi]y$.

Let φ be a functional postsign. In a similar fashion we say that

$$s \, \varepsilon \, \text{K} . \varphi \, \varepsilon \, s'\text{F} : \supset \, \therefore \, y[\varphi] = [x\varepsilon](x\varphi = y).$$

The sign [ ], of which we have already given several uses in logic, is called the sign of the inverse, for if α is a proposition containing the indeterminate x, and a is the class composed of those individuals which satisfy condition a, we have $x \, \varepsilon \, a . = . \alpha$, so that $a = [x\varepsilon]\alpha$, as in v, 1.

Let α be a formula containing the indeterminate x, and let φ be the functional presign which when placed before the letter x produces the formula α, that is, let $\alpha = \varphi x$. Then we have $\varphi = \alpha[x]$, and if x' is a new entity, we have $\varphi x' = \alpha[x]x'$. That is, if α is a formula containing the indeterminate x, then $\alpha[x]x'$ indicates what we obtain by putting x' in place of x.

Similarly, let α be a formula containing the indeterminate x, and let φ be a functional postsign, so that $x\varphi = \alpha$, and hence $\varphi = [x]\alpha$. Then, if x' is a new entity, we have $x'\varphi = x'[x]\alpha$, that is, $x'[x]\alpha$ again indicates what is obtained from α by reading x' in place of x, as in v, 4.

The sign [ ] can have other uses also in logic, which we present briefly, although we do not use them. Let a and b be two classes; then $[a \cap]b$ or $b[\cap a]$ indicates the classes of x which satisfy the condition $b = a \cap x$, or $b = x \cap a$. If b is not contained in a, no class satisfies this condition; if b is contained in a, the sign $b[\cap a]$ indicates all classes which contain b and are contained in $b \cup - a$.

In arithmetic, let a, b be numbers; then $b[+ a]$ or $[a +]b$ indicates the number x which satisfies the condition $b = x + a$ or $b = a + x$. Similarly we have $b[\times a] = [a \times]b = b/a$. And this sign can be used in analysis:

$$y = \sin x . = . x \,\varepsilon\, [\sin]y \qquad \text{(in place of } x = \text{ arc sin } y),$$
$$dF(x) = f(x)dx . = . F(x) \,\varepsilon\, [d]f(x)dx \quad \text{(in place of } F(x) = \int f(x)dx).$$

On the other hand, let φ be a functional presign on the class s, and let k be a class contained in s; then φk indicates the class made up of all φx, where the x are entities of the class k. That is:

DEF. $s\,\varepsilon\, K . k\,\varepsilon\, K . k \supset s . \varphi\,\varepsilon\, F's : \supset . \varphi k = [y\varepsilon](k . [\varphi]y : = -\Lambda)$.

Or $\quad s\,\varepsilon\, K . k\,\varepsilon\, K . k \supset s . \varphi\,\varepsilon\, F's : \supset . \varphi k = [y\varepsilon]([x\varepsilon] : x\,\varepsilon\, k . \varphi x = y$
$$\therefore - = \Lambda).$$

DEF. $s\,\varepsilon\, K . k\,\varepsilon\, K . k \supset s . \varphi\,\varepsilon\, s'F : \supset . k\varphi = [y\varepsilon](k . y[\varphi] : - = \Lambda)$.

So that, if $\varphi\,\varepsilon\, F's$, then φs indicates the class made up of all φx, where the x are entities of the class s. We then have:

$s\,\varepsilon\, K . \varphi\,\varepsilon\, F's . y\,\varepsilon\, \varphi s : \supset : \varphi[\varphi]y = y$.

$s\,\varepsilon\, K . a, b\,\varepsilon\, K . a \supset s . b \supset s . \varphi\,\varepsilon\, F's : \supset . \varphi(a \cup b) = (\varphi a) \cup (\varphi b)$.

$s\,\varepsilon\, K . \varphi\,\varepsilon\, F's : \supset . \varphi \Lambda = \Lambda$.

$s\,\varepsilon\, K . a, b\,\varepsilon\, K . b \supset s . a \supset b . \varphi\,\varepsilon\, F's : \supset . \varphi a \supset \varphi b$.

$s\,\varepsilon\, K . a, b\,\varepsilon\, K . a \supset s . b \supset s . \varphi\,\varepsilon\, F's : \supset . \varphi(ab) \supset (\varphi a)(\varphi b)$.

Let a be a class; then $a \cap K$, or $K \cap a$, or Ka, indicates all classes of the form $a \cap x$, or $x \cap a$, or xa, where x is any class; that is, Ka indicates the classes which are contained in a. The formula $x\,\varepsilon\, Ka$ means the same as $x\,\varepsilon\, K . x \supset a$. We shall sometimes use this convention; thus KN denotes the *class of numbers.*

Similarly, if a is a class, $K \cup a$ indicates the classes which contain a.

Let a be a number; then $a + N$, or $N + a$, indicates the *numbers larger than the number a*; $a \times$ N, or N $\times$ a, or Na, indicates the *multiples of the number a*; N^2, N^3, ... indicate *square numbers, cube numbers*, etc.

The equality of functional signs, their products, and powers may be defined as follows:

DEF. $s \, \varepsilon \, K \, . \, \varphi, \psi \, \varepsilon \, F's : \supset \, \therefore \, \varphi = \psi : = : x \, \varepsilon \, s \, . \, \supset_x \, . \, \varphi x = \psi x.$

DEF. $s \, \varepsilon \, K \, . \, \varphi \, \varepsilon \, F's \, . \, \psi \, \varepsilon \, F'\varphi s, x \, \varepsilon \, s : \supset \, . \, \psi \varphi x = \psi(\varphi x).$

Thus, on the assumption of this definition, $\psi \varphi$ is a new functional presign; and this is called the *product of the signs* ψ *and* φ.

Similarly, if φ, ψ are functional postsigns.

The following proposition holds:

$$s \, \varepsilon \, K \, . \, \varphi \, \varepsilon \, F's \, . \, \varphi s \supset s : \supset : \varphi \varphi s \supset s \, . \, \varphi \varphi \varphi s \supset s \, . \, \text{etc.}$$

The functions $\varphi \varphi$, $\varphi \varphi \varphi$, ... are called iterates, and are commonly indicated by the signs φ^2, φ^3, ..., as if they were powers of the operation φ.

But if φ is a functional postsign, we may use this easier notation, there being no ambiguity:

DEF. $s \, \varepsilon \, K \, . \, \varphi \, \varepsilon \, s'F \, . \, s\varphi \supset s : \supset : \varphi 1 = \varphi \, . \, \varphi 2 = \varphi \varphi \, . \, \varphi 3 = \varphi \varphi \varphi \, . \, \text{etc.}$

On the assumption of this definition, if $m, n \, \varepsilon$ N, we have $\varphi(m + n) = (\varphi m)(\varphi n); (\varphi m)n = \varphi(mn).$

If we use this definition in arithmetic, we find the following. We can indicate the number which follows the number a by the more suitable sign $a+$, and then $a + 1, a + 2$, ..., and $a + b$, if b is a number, have the meanings $a+, a++$, ..., as is clear from the definitions of §1. We can write proposition 6 of §1 as N$+ \supset$ N. If a, b, c are numbers, then $a : +b \, . \, c$ means $a + bc$, and $a : \times b \, . \, c$ means ab^c.

Functional signs enjoy many other properties, especially if they satisfy the condition $\varphi x = \varphi y \, . \, \supset \, . \, x = y$. A functional sign which satisfies this condition is called *similar* (*ähnliche Abbildung*) by Professor Dedekind. But space is lacking to present them here.

REMARKS

Definition, or briefly *Def.*, is a proposition having the form $x = a$, or $\alpha \supset \, . \, x = a$, where a is an aggregate of signs having a known meaning, x is a sign, or aggregate of signs, which up to this point lacks a meaning, and α is the context in which the definition is given.

Theorem (Theor. or Th) is a proposition that is proved. If a theorem has the form $\alpha \supset \beta$, where α and β are propositions, then α is said to be the *Hypothesis* (Hyp. or briefly Hp) and β the *Thesis* (Thes. or Ts). The Hyp.

and the Ts depend on the form of the theorem, for if we write $-\beta \supset -\alpha$ in place of $\alpha \supset \beta$, then $-\beta$ will be the Hp and $-\alpha$ will be the Ts, or if we write $\alpha - \beta = \Lambda$, then the Hp and the Ts are missing.

In every section, the sign P followed by a certain number indicates the proposition of that number in the same section. The propositions of logic are indicated by the sign L and the number of the proposition.

Formulas that are not contained on one line continue on the next with no intervening sign.

THE PRINCIPLES OF ARITHMETIC

1 NUMBERS AND ADDITION

Explanations

The sign N means *number* (*positive integer*); 1 means *unity*; $a + 1$ means the *successor of a*, or *a plus* 1; and $=$ means *is equal to* (this must be considered as a new sign, although it has the appearance of a sign of logic).

Axioms

1. $1 \, \varepsilon \, N.$
2. $a \, \varepsilon \, N . \supset . a = a.$
3. $a, b \, \varepsilon \, N . \supset : a = b . = . b = a.$
4. $a, b, c \, \varepsilon \, N . \supset \therefore a = b . b = c : \supset . a = c.$
5. $a = b . b \, \varepsilon \, N : \supset . a \, \varepsilon \, N.$
6. $a \, \varepsilon \, N . \supset . a + 1 \, \varepsilon \, N.$
7. $a, b \, \varepsilon \, N . \supset : a = b . = . a + 1 = b + 1.$
8. $a \, \varepsilon \, N . \supset . a + 1 - = 1.$
9. $k \, \varepsilon \, K \therefore 1 \, \varepsilon \, k \therefore x \, \varepsilon \, N . x \, \varepsilon \, k : \supset_x . x + 1 \, \varepsilon \, k :: \supset . N \supset k.$

Definitions

10. $2 = 1 + 1; 3 = 2 + 1; 4 = 3 + 1;$ etc.

Theorems

11. $2 \, \varepsilon \, N.$

PROOF

P1 . ɔ :	1 ε N	(1)
1[a](P6) . ɔ :	1 ε N . ɔ . 1 + 1 ε N	(2)
(1)(2) . ɔ :	1 + 1 ε N	(3)
P10 . ɔ :	2 = 1 + 1	(4)
(4) . (3) . (2, 1 + 1)[a, b](P5) : ɔ :	2 ε N	(Theorem)

NOTE We have explicitly written every step of this very easy proof. For the sake of brevity, we shall write it as follows:

P1 . 1[a](P6) : ɔ : 1 + 1 ε N . P10 . (2, 1 + 1)[a, b](P5) : ɔ : Th.

or P1 . P6 : ɔ : 1 + 1 ε N . P10 . P5 : ɔ : Th.

12. 3, 4, ... ε N.

13. a, b, c, d ε N . $a = b$. $b = c$. $c = d$: ɔ : $a = d$.

PROOF Hyp. P4 : ɔ : a, c, d ε N . $a = c$. $c = d$. P4 : ɔ : Thes.

14. a, b, c ε N . $a = b$. $b = c$. $a - = c$: = . Λ.

PROOF P4 . L39 : ɔ . Theor.

15. a, b, c ε N . $a = b$. $b - = c$: ɔ . $a - = c$.

16. a, b ε N . $a = b$: ɔ . $a + 1 = b + 1$.

16'. a, b ε N . $a + 1 = b + 1$: ɔ . $a = b$.

PROOF P7 = (P16)(P16').

17. a, b ε N . ɔ : $a - = b$. = . $a + 1 - = b + 1$.

PROOF P7 . L21 : ɔ . Theor.

Definition

18. a, b ε N . ɔ . $a + (b + 1) = (a + b) + 1$.

NOTE This definition should be read: if a and b are numbers, and $(a + b) + 1$ has meaning (that is, if $a + b$ is a number), but $a + (b + 1)$ has not yet been defined, then $a + (b + 1)$ indicates the number that follows $a + b$.

From this definition, and the preceding, we deduce that

a ε N . ɔ ∴ $a + 2 = a + (1 + 1) = (a + 1) + 1$,

a ε N . ɔ ∴ $a + 3 = a + (2 + 1) = (a + 2) + 1$, etc.

Theorems

19. $a, b \, \varepsilon \, N . \supset . a + b \, \varepsilon \, N.$

PROOF

$a \, \varepsilon \, N . P6 : \supset : a + 1 \, \varepsilon \, N : \supset : 1 \, \varepsilon \, [b\varepsilon]Ts.$ (1)

$a \, \varepsilon \, N . \supset :: b \, \varepsilon \, N . b \, \varepsilon \, [b\varepsilon]Ts : \supset : a + b \, \varepsilon \, N . P6 : \supset : (a + b) +$
$1 \, \varepsilon \, N . P18 : \supset : a + (b + 1) \, \varepsilon \, N : \supset : (b + 1) \, \varepsilon \, [b\varepsilon]Ts.$ (2)

$a \, \varepsilon \, N . (1) . (2) . \supset :: 1 \, \varepsilon \, [b\varepsilon]Ts \therefore b \, \varepsilon \, N . b \, \varepsilon \, [b\varepsilon]Ts : \supset :$
$b + 1 \, \varepsilon \, [b\varepsilon]Ts \therefore ([b\varepsilon]Ts)[k]P9 :: \supset : N \supset [b\varepsilon]Ts. (L50) ::$
$\supset : b \, \varepsilon \, N . \supset . Ts.$ (3)

$(3) . (L42) : \supset : a, b \, \varepsilon \, N . \supset .$ Thesis.   (Theor.)

20. DEF. $a + b + c = (a + b) + c.$

21. $a, b, c \, \varepsilon \, N . \supset . a + b + c \, \varepsilon \, N.$

22. $a, b, c \, \varepsilon \, N . \supset : a = b . = . a + c = b + c.$

PROOF

$a, b \, \varepsilon \, N . P7 : \supset . 1 \, \varepsilon \, [c\varepsilon]Ts.$ (1)

$a, b \, \varepsilon \, N . \supset :: c \, \varepsilon \, N . c \, \varepsilon \, [c\varepsilon]Ts \therefore \supset \therefore a = b . = . a + c = b + c :$
$a + c, b + c \, \varepsilon \, N : a + c = b + c . = . a + c + 1 = b + c + 1$
$\therefore \supset \therefore a = b . = . a + (c + 1) = b + (c + 1) \therefore \supset \therefore$
$(c + 1) \, \varepsilon \, [c\varepsilon]Ts.$ (2)

$a, b \, \varepsilon \, N . (1) . (2) : \supset :: 1 \, \varepsilon \, [c\varepsilon]Ts \therefore c \, \varepsilon \, [c\varepsilon]Ts . \supset . (c + 1) \, \varepsilon \, [c\varepsilon]Ts$
$:: \supset :: c \, \varepsilon \, N . \supset . Ts.$ (3)

$(3) \supset$ Theor.

23. $a, b, c \, \varepsilon \, N . \supset . a + (b + c) = a + b + c.$

PROOF

$a, b \, \varepsilon \, N . P18 . P20 : \supset . 1 \, \varepsilon \, [c\varepsilon]Ts.$ (1)

$a, b \, \varepsilon \, N . \supset \therefore c \, \varepsilon \, N . c \, \varepsilon \, [c\varepsilon]Ts : \supset : a + (b + c) = a + b + c.$
$P7 : \supset : a + (b + c) + 1 = a + b + c + 1 . P18 : \supset :$
$a + (b + (c + 1)) = a + b + (c + 1) : \supset . c + 1 \, \varepsilon \, [c\varepsilon]Ts.$ (2)

$(1)(2)(P9) . \supset .$ Theor.

24. $a \, \varepsilon \, N . \supset . 1 + a = a + 1.$

PROOF

P2 . ⊃ . 1 ε [aε]Ts.   (1)

a ε N . a ε [aε]Ts : ⊃ : 1 + a = a + 1 : ⊃ : 1 + (a + 1) =
(a + 1) + 1 : ⊃ : (a + 1) ε [aε]Ts.   (2)

(1)(2) . ⊃ . Theor.

24′. a, b ε N . ⊃ . 1 + a + b = a + 1 + b.

PROOF Hyp. P24 : ⊃ : 1 + a = a + 1 . P22 : ⊃ . Thesis.

25. a, b ε N . ⊃ . a + b = b + a.

PROOF

a ε N . P24 : ⊃ : 1 ε [bε]Ts.   (1)

a ε N . ⊃ ∴ b ε N . b ε [bε]Ts : ⊃ : a + b = b + a . P7 : ⊃ :
(a + b) + 1 = (b + a) + 1 . (a + b) + 1 = a + (b + 1) .
(b + a) + 1 = 1 + (b + a) . 1 + (b + a) = (1 + b) + a .
(1 + b) + a = (b + 1) + a : ⊃ : a + (b + 1) =
(b + 1) + a : ⊃ : (b + 1) ε [bε]Ts.   (2)

(1)(2) . ⊃ . Theor.

26. a, b, c ε N . ⊃ : a = b . = . c + a = c + b.

27. a, b, c ε N . ⊃ : a + b + c = a + c + b.

28. a, b, c, d ε N . a = b . c = d : ⊃ . a + c = b + d.

2 SUBTRACTION

Explanations

The sign − is read *minus*, < is read *is less than*, and > is read *is greater than*.

Definitions

1. a, b ε N . ⊃ : b − a = N[xε](x + a = b).
2. a, b ε N . ⊃ : a < b . = . b − a − = Λ.
3. a, b ε N . ⊃ : b > a . = . a < b.
   a + b − c = (a + b) − c; a − b + c = (a − b) + c;
   a − b − c = (a − b) − c.

Theorems

4. $a, b, a', b' \, \varepsilon \, N \, . \, a = a' \, . \, b = b' \, : \supset \, : b - a = b' - a'.$

PROOF Hyp . $\supset : x + a = b \, . \, = \, . \, x + a' = b' : \supset$. Thesis.

5. $a, b \, \varepsilon \, N \, . \supset \, : a < b \, . \, = \, . \, b - a \, \varepsilon \, N.$

PROOF

$a, b \, \varepsilon \, N : \supset \, \therefore \, x, y \, \varepsilon \, b - a \, . \supset_{x,y} : x, y \, \varepsilon \, N \, . \, x + a = b \, .$
$y + a = b \, . \, \S 1P22 : \supset : x = y.$ (1)
$a, b \, \varepsilon \, N \, . \, a < b \, . \, P2 \, . \, (1) : \supset \, \therefore \, b - a - = \Lambda : x, y \, \varepsilon \, b - a \, .$
$\supset \, . \, x = y : (N, b - a)[s, k](L56) \, \therefore \, \supset \, \therefore \, b - a \, \varepsilon \, N.$ (2)
$a, b \, \varepsilon \, N \, . \, b - a \, \varepsilon \, N \, . \, (L56) : \supset : b - a - = \Lambda : \supset : a < b.$ (3)
$(2)(3) \, . \supset$. Theor.

6. $a, b \, \varepsilon \, N \, . \, a < b : \supset \, . \, b - a + a = b.$

PROOF Hyp . P5 . P1 $: \supset : b - a \, \varepsilon \, N \, . \, (b - a) \, \varepsilon \, [x \varepsilon](x + a = b) : \supset :$
Thes.

7. $a, b, c \, \varepsilon \, N \, . \supset \, : c = b - a \, . \, = \, . \, c + a = b.$

PROOF Hyp . $\S 1P22 \, . \, P6 : \supset : c = b - a \, . \, = \, . \, c + a = b - a + a \, . \, =$
$. \, c + a = b.$

8. $a, b \, \varepsilon \, N \, . \supset \, . \, a + b - a = b.$

PROOF $(a + b, b)[b, c]P7 \, . \supset$. Theor.

9. $a, b, c \, \varepsilon \, N \, . \, a < b : \supset : c + (b - a) = c + b - a.$

PROOF Hyp . P6 $: \supset : (b - a) + a = b : \supset : c + (b - a) + a =$
$c + b \, . \, P7 : \supset :$ Thesis.

10. $a, b, c \, \varepsilon \, N \, . \, a > b + c : \supset \, . \, a - (b + c) = a - b - c.$

11. $a, b, c \, \varepsilon \, N \, . \, b > c \, . \, a > b - c : \supset \, . \, a - (b - c) = a + c - b.$

12. $a, b, a', b' \, \varepsilon \, N \, . \, a = a' \, . \, b = b' : \supset : a < b \, . \, = \, . \, a' < b'.$

PROOF Hyp . $\supset \, . \, b - a = b' - a' \, . \supset \, . \, b - a \, \varepsilon \, N = b' - a' \, \varepsilon \, N \, . \supset .$
Thes.

13. $a, b \, \varepsilon \, N \, . \supset \, . \, a < a + b.$

PROOF Hyp . P8 $: \supset : a + b - a = b : \supset \, . \, a + b - a \, \varepsilon \, N \, . \, P5 : \supset :$
Thesis.

14. $a, b, c \, \varepsilon \, N . a < b . b < c : \supset . a < c.$

PROOF Hyp $. \supset : b - a \, \varepsilon \, N . c - b \, \varepsilon \, N : \supset : (b - a) + (c - b) \, \varepsilon \, N :$ $\supset : c - a \, \varepsilon \, N : \supset .$ Thesis.

15. $a, b, c \, \varepsilon \, N . \supset : a < b . = . a + c < b + c.$

PROOF Hyp $. \supset : a < b . = . b - a \, \varepsilon \, N . = . (b + c) - (a + c) \, \varepsilon \, N . =$ $. a + c < b + c.$

16. $a, b, a', b' \, \varepsilon \, N . a < b . a' < b' : \supset . a + a' < b + b'.$

PROOF Hyp $. \supset : a + a' < b + a' . b + a' < b + b' : \supset .$ Thesis.

17. $a, b, c \, \varepsilon \, N . a < b < c : \supset . c - a > c - b.$

PROOF Hyp $. \supset . b - a \, \varepsilon \, N . c - b \, \varepsilon \, N . (c - b) + (b - a) = c - a : \supset .$ Thesis.

18. $a \, \varepsilon \, N . \supset : a = 1 . \cup . a > 1.$

PROOF

$1 \, \varepsilon \, [a\varepsilon]$Thesis. (1)

$a \, \varepsilon \, N . P13 : \supset : a + 1 > 1 : \supset : a + 1 \, \varepsilon \, [a\varepsilon]$Thesis. (2)

$(1)(2) . \supset .$ Theor.

19. $a, b \, \varepsilon \, N . \supset . a + b - = b.$

PROOF

$a \, \varepsilon \, N . \S 1P8 : \supset : a + 1 - = 1 : \supset : 1 \, \varepsilon \, [b\varepsilon]$Thesis. (1)

$a \, \varepsilon \, N . b \, \varepsilon \, N . b \, \varepsilon \, [b\varepsilon]$Ts $: \supset : a + b - = b . \S 1P17 : \supset :$ $a + (b + 1) - = b + 1 : \supset : b + 1 \, \varepsilon \, [b\varepsilon]$Ts. (2)

$(1)(2) . \supset .$ Theor.

20. $a, b \, \varepsilon \, N . a < b . a = b : = . \Lambda.$

PROOF Hyp $: \supset : b - a \, \varepsilon \, N . (b - a) + a = a . P19 : \supset : \Lambda.$

21. $a, b \, \varepsilon \, N . a > b . a = b : = . \Lambda.$

22. $a, b \, \varepsilon \, N . a > b . a < b : = . \Lambda.$

23. $a, b \, \varepsilon \, N : \supset : a < b . \cup . a = b . \cup . a > b.$

PROOF

$a \, \varepsilon \, N . P18 : \supset . 1 \, \varepsilon \, [b\varepsilon]$Ts. (1)

$a, b \, \varepsilon \, N . a < b : \supset . a < b + 1.$ (2)

$a, b \, \varepsilon \, N . a = b : \supset . a < b + 1.$ (3)

$a, b \, \varepsilon \, N . a > b : \supset : a - b \, \varepsilon \, N . P18 : \supset :$

$\quad a - b = 1 . \cup . a - b > 1.$ \hfill (4)

$a, b \, \varepsilon \, N . a - b = 1 : \supset . a = b + 1.$ \hfill (5)

$a, b \, \varepsilon \, N . a - b > 1 : \supset . a > b + 1.$ \hfill (6)

$a, b \, \varepsilon \, N . a > b . (4)(5)(6) : \supset : a = b + 1 . \cup . a > b + 1.$ \hfill (7)

$a, b \, \varepsilon \, N : a < b . \cup . a = b . \cup . a > b : (2)(3)(7) \therefore \supset \therefore$

$\quad a < b + 1 . \cup . a = b + 1 . \cup . a > b + 1.$ \hfill (8)

$a, b \, \varepsilon \, N . b \, \varepsilon \, [b\varepsilon]Ts. (8) : \supset : b + 1 \, \varepsilon \, [b\varepsilon]Ts.$ \hfill (9)

$(1)(9) . \supset .$ Theor.

3 MAXIMA AND MINIMA

Explanations

Let $a \, \varepsilon \, KN$, that is, let a be a class of numbers; then Ma is read *greatest among a*, and Wa is read *least among* a.

Definitions

1. $a \, \varepsilon \, KN . \supset . Ma = [x\varepsilon](x \, \varepsilon \, a \therefore a . \mathfrak{z} > x : = \Lambda).$

2. $a \, \varepsilon \, KN . \supset . Wa = [x\varepsilon](x \, \varepsilon \, a \therefore a . \mathfrak{z} < x : = \Lambda).$

3. $n \, \varepsilon \, N . a \, \varepsilon \, KN . a - = \Lambda . a\mathfrak{z} > n = \Lambda : \supset . Ma \, \varepsilon \, N.$

PROOF

$a \, \varepsilon \, KN . a - = \Lambda . a\mathfrak{z} > 1 = \Lambda : \supset : a = 1 : \supset . Ma = 1 :$

$\quad \supset . Ma \, \varepsilon \, N.$ \hfill (1)

$(1)\supset : 1 \, \varepsilon \, [n\varepsilon](Hp \supset Ts).$ \hfill (2)

$n \, \varepsilon \, N . a \, \varepsilon \, KN . a\mathfrak{z} > n + 1 = \Lambda . n + 1 \, \varepsilon \, a : \supset : n + 1 = Ma :$

$\quad \supset : Ma \, \varepsilon \, N.$ \hfill (3)

$n \, \varepsilon \, N . a \, \varepsilon \, KN . a\mathfrak{z} > n + 1 = \Lambda . n + 1 - \varepsilon \, a :$

$\quad \supset : a\mathfrak{z} > n = \Lambda.$ \hfill (4)

$n \, \varepsilon \, [n\varepsilon](Hp \supset Ts) . a \, \varepsilon \, KN . a\mathfrak{z} > n + 1 = \Lambda . n + 1 - \varepsilon \, a :$

$\quad \supset : Ma \, \varepsilon \, N.$ \hfill (5)

$n \, \varepsilon \, [n\varepsilon](Hp \supset Ts) . a \, \varepsilon \, KN . a\mathfrak{z} > n + 1 = \Lambda . (3)(5) :$

$\quad \supset : Ma \, \varepsilon \, N.$ \hfill (6)

$n \, \varepsilon \, [n\varepsilon](Hp \supset Ts) . (6) : \supset . (n + 1) \, \varepsilon \, [n\varepsilon](Hp \supset Ts).$ \hfill (7)

$(2)(7) . \S 1P9 : \supset : n \, \varepsilon \, N . \supset . Hp \supset Ts$ \hfill (Theor.)

4. $a \varepsilon \text{KN} . a - = \Lambda : \supset . \text{W}a \varepsilon \text{N}.$

5. $a \varepsilon \text{KN} . \supset . \text{W}a = \text{M}[x\varepsilon](a_3 < x = \Lambda).$

4 MULTIPLICATION

Definitions

1. $a \varepsilon \text{N} . \supset . a \times 1 = a.$
2. $a, b \varepsilon \text{N} . \supset . a \times (b + 1) = a \times b + a . ab = a \times b;$
   $ab + c = (ab) + c; abc = (ab)c.$

Theorems

3. $a, b \varepsilon \text{N} . \supset . ab \varepsilon \text{N}.$

PROOF

$a \varepsilon \text{N} . \text{P1} : \supset : a \times 1 \varepsilon \text{N} : \supset . 1 \varepsilon [b\varepsilon]\text{Ts}.$ (1)

$a, b \varepsilon \text{N} . b \varepsilon [b\varepsilon]\text{Ts} : \supset : a \times b \varepsilon \text{N} . \S1\text{P19} : \supset : ab + a \varepsilon \text{N}.$

$\text{P1} : \supset : a(b + 1) \varepsilon \text{N} : \supset : b + 1 \varepsilon [b\varepsilon]\text{Ts}.$ (2)

$(1)(2) . \supset . \text{Theor}.$

4. $a, b, c \varepsilon \text{N} . \supset . (a + b)c = ac + bc.$

NOTE This is the fifth proposition of Euclid's *Elements*, Book VII.

PROOF

$a, b \varepsilon \text{N} . \text{P1} : \supset : 1 \varepsilon [c\varepsilon]\text{Ts}.$ (1)

$a, b, c \varepsilon \text{N} . c \varepsilon [c\varepsilon]\text{Ts} : \supset : (a + b)c = ac + bc . \S1\text{P22} : \supset :$

$(a + b)c + a + b = ac + bc + a + b . \text{P2} : \supset :$

$(a + b)(c + 1) = a(c + 1) + b(c + 1) : \supset : c + 1 \varepsilon [c\varepsilon]\text{Ts}.$ (2)

$(1)(2) . \supset . \text{Theor}.$

5. $a \varepsilon \text{N} . \supset . 1 \times a = a.$

PROOF

$1 \varepsilon [a\varepsilon]\text{Ts}.$ (1)

$a \varepsilon [a\varepsilon]\text{Ts} . \supset . 1 \times a = a . \supset . 1 \times a + 1 = a + 1 .$

$\supset . 1 \times (a + 1) = a + 1 . \supset . a + 1 \varepsilon [a\varepsilon]\text{Ts}.$ (2)

$(1)(2) . \supset . \text{Theor}.$

6. $a, b \varepsilon \text{N} . \supset . ba + a = (b + 1)a.$

7. $a, b \varepsilon \text{N} . \supset . ab = ba.$ (Euclid, VII, 16)

PROOF

$a \, \varepsilon \, \text{N} \,.\, \text{P5} \,.\, \text{P1} : \supset .\, a \times 1 = a = 1 \times a : \supset : 1 \, \varepsilon \, [b\varepsilon]\text{Ts}.$ (1)

$a, b \, \varepsilon \, \text{N} \,.\, b \, \varepsilon \, [b\varepsilon]\text{Ts} : \supset : ab = ba : \supset : ab + a = ba + a \,.$

    $\text{P1} \,.\, \text{P6} : \supset : a(b + 1) = (b + 1)a : \supset : b + 1 \, \varepsilon \, [b\varepsilon]\text{Ts}.$ (2)

(1)(2) . $\supset$. Theor.

8. $a, b, c \, \varepsilon \, \text{N} \,.\, \supset .\, a(b + c) = ab + ac.$

PROOF   P4 . P7 : $\supset$. Theor.

9. $a, b, c \, \varepsilon \, \text{N} \,.\, a = b : \supset : ac = bc.$

PROOF

$a, b \, \varepsilon \, \text{N} \,.\, a = b :: \supset :: 1 \, \varepsilon \, [c\varepsilon]\text{Ts} \,\therefore\, c \, \varepsilon \, [c\varepsilon]\text{Ts} \,.\, \supset : ac = bc \,.$

    $a = b : \supset : ac + a = bc + b : \supset : a(c + 1) = b(c + 1) :$

    $\supset : c + 1 \, \varepsilon \, [c\varepsilon]\text{Ts} :: \supset : c \, \varepsilon \, \text{N} \,.\, \supset \,.\, \text{Ts}.$

10. $a, b, c \, \varepsilon \, \text{N} \,.\, a < b : \supset .\, (b - a)c = bc - ac.$ (Euclid, VII, 7)

PROOF   Hyp . $\supset : b - a \, \varepsilon \, \text{N} \,.\, (b - a) + a = b : \supset : (b - a)c + ac = bc : \supset : (b - a)c = bc - ac.$

11. $a, b, c \, \varepsilon \, \text{N} \,.\, a < b : \supset : ac < bc.$

PROOF   Hyp . $\supset : b - a \, \varepsilon \, \text{N} \,.\, \text{P3} : \supset : (b - a)c \, \varepsilon \, \text{N} \,.\, \text{P10} : \supset : bc - ac \, \varepsilon \, \text{N} : \supset$ Thesis.

12. $a, b, c \, \varepsilon \, \text{N} \,.\, \supset \,\therefore\, a < b \,.\, = .\, ac < bc : a = b \,.\, = .\, ac = bc : a > b \,.\, = .\, ac > bc.$

13. $a, b, a', b' \, \varepsilon \, \text{N} \,.\, a < a' \,.\, b < b' : \supset : ab < a'b'.$

14. $a, b \, \varepsilon \, \text{N} : \supset : ab \,.\, > \cup = .\, a.$

15. $a, b, c \, \varepsilon \, \text{N} \,.\, \supset .\, a(bc) = abc.$

PROOF

$a, b \, \varepsilon \, \text{N} \,.\, \text{P1} : \supset : 1 \, \varepsilon \, [c\varepsilon]\text{Ts}.$ (1)

$a, b, c \, \varepsilon \, \text{N} \,.\, c \, \varepsilon \, [c\varepsilon]\text{Ts} : \supset : a(bc) = abc : \supset : a(bc) + ab = abc + ab : \supset : a(bc + b) = ab(c + 1)$

    $: \supset : a(b(c + 1)) = ab(c + 1) : \supset : c + 1 \, \varepsilon \, [c\varepsilon]\text{Ts}.$ (2)

(1)(2) . $\supset$. Theor.

5 POWERS

Definitions

1. $a \, \varepsilon \, N . \supset . a^1 = a.$
2. $a, b \, \varepsilon \, N . \supset . a^{b+1} = a^b a.$

Theorems

3. $a, b \, \varepsilon \, N . \supset . a^b \, \varepsilon \, N.$

PROOF

$a \, \varepsilon \, N . P1 : \supset . 1 \, \varepsilon \, [b\varepsilon]Ts.$ (1)

$a, b \, \varepsilon \, N . b \, \varepsilon \, [b\varepsilon]Ts : \supset : a^b \, \varepsilon \, N . §4P3 : \supset : a^b a \, \varepsilon \, N . P1$

$\quad : \supset : a^{b+1} \, \varepsilon \, N : \supset : b + 1 \, \varepsilon \, [b\varepsilon]Ts.$ (2)

$(1)(2) . \supset . \text{Theor.}$

4. $a \, \varepsilon \, N . \supset . 1^a = 1.$
5. $a, b, c \, \varepsilon \, N . \supset . a^{b+c} = a^b a^c.$
6. $a, b, c \, \varepsilon \, N . \supset . (ab)^c = a^c b^c.$
7. $a, b, c \, \varepsilon \, N . \supset . (a^b)^c = a^{bc}.$
8. $a, b, c \, \varepsilon \, N . \supset \therefore a < b . = . a^c < b^c :$
   $a = b . = . a^c = b^c : a > b . = . a^c > b^c.$
9. $a, b, c \, \varepsilon \, N . a > 1 . \supset \therefore b < c . = . a^b < a^c :$
   $b = c . = . a^b = a^c : b > c . = . a^b > a^c.$

6 DIVISION

Explanations

The sign / is read *divided by*, D is read *divides*, or *is a divisor of*, ꓷ is read *is a multiple of*, Np is read *prime number*, and π is read *is prime with*.

Definitions

1. $a, b \, \varepsilon \, N . \supset . b / a = N[x\varepsilon](xa = b).$
2. $a, b \, \varepsilon \, N . \supset : a \, D \, b . = . b / a - = \Lambda.$
3. $a, b \, \varepsilon \, N . \supset : b \, ꓷ \, a . = . a \, D \, b.$
4. $Np = N[x\varepsilon](3 \, D \, x . 3 > 1 . 3 < x : = \Lambda).$
5. $a, b \, \varepsilon \, N . \supset :: a \, \pi \, b \therefore = \therefore 3 \, D \, a . 3 \, D \, b . 3 > 1 : = \Lambda.$
6. $a, b \, \varepsilon \, N . \supset \therefore 3 \, D \, (a, b) : = : 3 \, D \, a . \cap . 3 \, D \, b.$

7. $a, b \, \varepsilon \, N . \supset \therefore \, 3 \, \sqcap (a, b) : = \, : 3 \, \sqcap a . \cap . 3 \, \sqcap b . ab/c = (ab)/c;$
$a/b/c = (a/b)/c; a/b \times c = (a/b)c.$

Theorems

NOTE Theorems 8–30 are proved as for subtraction.

8. $a, b, a', b' \, \varepsilon \, N . a = a' . b = b' : \supset . a/b = a'/b'.$

9. $a, b, a', b' \, \varepsilon \, N . a = a' . b = b' : \supset : a \, D \, b . = . a' \, D \, b'.$

10. $a, b, c \, \varepsilon \, N . \supset : ac = b . = . c = b/a.$

11. $a, b \, \varepsilon \, N . \supset : a \, D \, b . = . b/a \, \varepsilon \, N.$

12. $a \, \varepsilon \, N . \supset . a/1 = a.$

13. $a \, \varepsilon \, N . \supset . a/a = 1.$

14. $a \, \varepsilon \, N . \supset . 1 \, D \, a.$

15. $a \, \varepsilon \, N . \supset . a \, D \, a.$

16. $a, b \, \varepsilon \, N . \supset . ab/b = a.$

17. $a, b \, \varepsilon \, N . a \, D \, b : \supset . a(b/a) = b.$

18. $a, b, c \, \varepsilon \, N . c \, D \, b : \supset . a(b/c) = ab/c.$

19. $a, b, c \, \varepsilon \, N . a \, \sqcap \, bc : \supset : a/(bc) = a/b/c.$

20. $a, b, c \, \varepsilon \, N . a \, \sqcap \, b . b \, \sqcap \, c : \supset . a/(b/c) = a/b \times c.$

21. $a, m, n \, \varepsilon \, N . m > n : \supset . a^m/a^n = a^{m-n}.$

22. $a, b \, \varepsilon \, N . \supset . a \, D \, ab.$

23. $a, b, c \, \varepsilon \, N . a \, D \, b . b \, D \, c : \supset . a \, D \, c.$

24. $a, b, c \, \varepsilon \, N . a \, D \, b . b \, D \, c : \supset . c/a \, \sqcap \, c/b.$

25. $a, b, c \, \varepsilon \, N . c \, D \, a . c \, D \, b : \supset . (a + b)/c = a/c + b/c.$

26. $a, b, c \, \varepsilon \, N . c \, D \, a . c \, D \, b . a > b : \supset : (a - b)/c = a/c - b/c.$

27. $a, b, c \, \varepsilon \, N . c \, D \, a . c \, D \, b : \supset . c \, D \, a + b.$

28. $a, b, c \, \varepsilon \, N . c \, D \, a . c \, D \, b . a > b : \supset . c \, D \, a - b.$

29. $a, b, c, m, n \, \varepsilon \, N . c \, D \, a . c \, D \, b : \supset . c \, D \, ma + nb.$

30. $a, b, c, m, n \, \varepsilon \, N . c \, D \, a . c \, D \, b . ma > nb : \supset . c \, D \, ma - nb.$

31. $a, b \, \varepsilon \, N . a \, D \, b : \supset : a . < \cup = . b.$

PROOF Hyp. P11 . P17 . §4P14 $: \supset : b/a \, \varepsilon \, N . a(b/a) = b . a < \cup =$
$a(b/a) : \supset .$ Thesis.

32. $a, b \, \varepsilon \, N . a \, D \, b . b \, D \, a : \supset . a = b.$

33. $a \, \varepsilon \, N . \supset . M \, 3 \, D \, a = a.$

34. $a, b \, \varepsilon \, N . a > b : \supset . 3 \, D \, (a, b) = 3 \, D \, (b, a - b).$

PROOF

  Hyp. P28 : ⊃ ∴ x D a . x D b : ⊃ : x D b . x D $(a - b)$.   (1)

  Hyp. P27 : ⊃ ∴ x D b . x D $(a - b)$

   : ⊃ : x D b . x D $(b + (a - b))$: ⊃ : x D b . x D a.   (2)

  (1)(2) ⊃ : Hyp . ⊃ ∴ x D a . x D b : = : x D b . x D $(a - b)$. (Theor.)

35. a, b ε N . ⊃ : M з D (a, b) ε N.

PROOF

  1 D a . 1 D b : ⊃ : з D (a, b) − = Λ.   (1)

  з D (a, b) . з > a : = Λ.   (2)

  (1)(2) . §3P3 : ⊃ . Th.

36. a, b ε N . ⊃ . з D (a, b) = з D M з D (a, b).   (Euclid, VII, 2)

PROOF

  k = N [cε](Hp. $a < c$. $b < c$: ⊃ . Ts.).   (1)

  a ε N . b ε N . $a < 1$. $b < 1$: = Λ.   (2)

  (1)(2) . ⊃ . 1 ε k.   (3)

  a, b ε N . $a < c + 1$. $b < c + 1$: ⊃ ∴ $a < c$. $b < c$: ∪ :

   $a = c$. $b < c$: ∪ : $a < c$. $b = c$: ∪ : $a = c$. $b = c$.   (4)

  c ε k . a, b ε N . $a < c$. $b < c$: ⊃ : Ts.   (5)

  c ε k . a, b ε N . $a = c$. $b < c$: ⊃ : c ε k . $b < c$. $a - b < c$.

   з D (a, b) = з D $(b, a - b)$: ⊃ : з D $(b, a - b)$ =

   з D M з D $(b, a - b)$: ⊃ : з D (a, b) =

   з D M з D (a, b) : ⊃ : Ts.   (6)

  $(a, b)[b, a](6)$ ⊃ . c ε k . a, b ε N . $a < c$. $b = c$: ⊃ : Ts.   (7)

  c ε k . a, b ε N . $a = c$. $b = c$: ⊃ : з D (a, b) = з D c =

   з D M з D c = з D M з D (a, b) : ⊃ : Ts.   (8)

  (4)(5)(6)(7)(8) . ⊃ . c ε k . a, b ε N .

   $a < c + 1$. $b < c + 1$: ⊃ : Ts.   (9)

  (9) ⊃ . c ε k . ⊃ . $(c + 1)$ ε k.   (10)

  (1)(10) . ⊃ ∴ c ε N . Hp . $a < c$. $b < c$: ⊃ : Ts.   (11)

  $(a + b)[c](11)$. ⊃ : Hp . ⊃ . Ts.   (Theor.)

37. a, b, m ε N . ⊃ . M з D (am, bm) = m × M з D (a, b).

7 VARIOUS THEOREMS

 1. a, b ε N . $a^2 + b^2$ ◖ 7 : ⊃ : a ◖ 7 . b ◖ 7.

 2. x ε N . ⊃ . $x(x + 1)$ ◖ 2.

3. $x \, \varepsilon \, N . \supset . x(x + 1)(x + 2) \, \complement \, 6.$

4. $x \, \varepsilon \, N . \supset . x(x + 1)(2x + 1) \, \complement \, 6.$

5. $x \, \varepsilon \, N . \supset : x . \pi . x + 1.$

6. $x \, \varepsilon \, N . \supset : 2x - 1 . \pi . 2x + 1.$

7. $x \, \varepsilon \, N . \supset . (2x + 1)^2 - 1 \, \complement \, 8.$

8. $a \, \varepsilon \, N . a > 1 : \supset \, \therefore \, Np . {}_3 > 1 . {}_3 \, D \, a : - = \Lambda.$ (Euclid, VII, 31)

9. $a, b \, \varepsilon \, N \, \therefore \, b^2 > a \, \therefore \, {}_3 \, D \, a . {}_3 > 1 . {}_3 < b : = \Lambda :: \supset . a \, \varepsilon \, Np.$

10. $a, b \, \varepsilon \, N . a \, \varepsilon \, Np . a - D \, b : \supset : a \, \pi \, b.$ (Euclid, VII, 29)

11. $a, b, c \, \varepsilon \, N . a \, D \, bc . a \, \pi \, b : \supset . a \, D \, c.$

12. $a, b \, \varepsilon \, N . m = M \, {}_3 \, D \, (a, b) : \supset : a/m . \pi . b/m.$

13. $a \, \varepsilon \, Np . b, c \, \varepsilon \, N . a \, D \, bc : \supset : a \, D \, b . \cup . a \, D \, c.$ (Euclid, VII, 30)

14. $a \, \varepsilon \, Np . b, n \, \varepsilon \, N : \supset : a \, D \, b^n . = . a \, D \, b.$ (Euclid, IX, 12)

15. $a, b, c \, \varepsilon \, N . a \, \pi \, b . c \, D \, a : \supset . c \, \pi \, b.$ (Euclid, VII, 23)

16. $a, b, c \, \varepsilon \, N . \supset \, \therefore \, a \, \pi \, b . a \, \pi \, c : = : a \, \pi \, bc.$ (Euclid, VII, 24)

17. $a, b, c \, \varepsilon \, N . b \, \pi \, c . b \, D \, a . c \, D \, a : \supset . bc \, D \, a.$

18. $a, b, c \, \varepsilon \, N . a \, \pi \, b : \supset : {}_3 \, D \, (ac, b) = {}_3 \, D \, (c, b).$

19. $a, b \, \varepsilon \, N . \supset . M \, {}_3 \, \complement \, (a, b) \, \varepsilon \, N.$

20. $a, b \, \varepsilon \, N . \supset . M \, {}_3 \, \complement \, (a, b) = ab/M \, {}_3 \, D \, (a, b).$ (Euclid, VII, 34)

21. $a, b, c \, \varepsilon \, N . c \, \complement \, a . c \, \complement \, b : \supset : c \, \complement \, M \, {}_3 \, \complement \, (a, b).$ (Euclid, VII, 35)

22. $x \, \varepsilon \, N . x < 41 : \supset . 41 - x + x^2 \, \varepsilon \, Np.$

23. $M . Np : = \Lambda.$ (Euclid, IX, 20)

24. $n \, \varepsilon \, Np . a \, \varepsilon \, N . a - \complement \, n : \supset . a^{n-1} - 1 \, \complement \, n.$ (Fermat)

8 RATIONAL NUMBERS

Explanations

If $p, q \, \varepsilon \, N$, then $\dfrac{p}{q}$ is read *the ratio of the number p to the number q.*

The sign R is read *ratio of two numbers*, and indicates the positive rational numbers.

Definitions

1. $m, p, q \, \varepsilon \, N . \supset . m \dfrac{p}{q} = mp/q.$

2. $p, q, p', q' \, \varepsilon \, N . \supset :: \dfrac{p}{q} = \dfrac{p'}{q'} . = \, \therefore \, x \, \varepsilon \, N . x \dfrac{p}{q}, x \dfrac{p'}{q'} \, \varepsilon \, N :$

$\quad \supset_x . x \dfrac{p}{q} = x \dfrac{p'}{q'} .$

3. $\;\; R = :: [x \, \varepsilon] \;\therefore\; p, q \, \varepsilon \, N . \dfrac{p}{q} = x : - =_{p,q} \Lambda.$

4. $\;\; p \, \varepsilon \, N . \supset . \dfrac{p}{1} = p.$

Theorems

5. $\;\; p, q, p', q' \, \varepsilon \, N . \supset :: \dfrac{p}{q} = \dfrac{p'}{q'} . = . pq' = p'q.$ (Euclid, VII, 19)

PROOF

$Hp . \dfrac{p}{q} = \dfrac{p'}{q'} : \supset \,\therefore\, qq', qq' \dfrac{p}{q}, qq' \dfrac{p'}{q'} \, \varepsilon \, N . \, P2 \,\therefore\, \supset \,\therefore\, qq' \dfrac{p}{q} = qq' \dfrac{p'}{q'}$

$. \, qq' \dfrac{p}{q} = pq' . \, qq' \dfrac{p'}{q'} = p'q \,\therefore\, \supset \,\therefore\, pq' = p'q.$ (1)

$Hp . pq' = p'q \,\therefore\, \supset \,\therefore\, x \, \varepsilon \, N . \, x \dfrac{p}{q}, x \dfrac{p'}{q'} \, \varepsilon \, N : \supset_x : xpq' = xp'q : \supset$

$: \left(x \dfrac{p}{q} \right) qq' = \left(x \dfrac{p'}{q'} \right) qq' : \supset : x \dfrac{p}{q} = x \dfrac{p'}{q'} .$ (2)

$(1)(2) . \supset . \text{Th.}$

6. $\;\; m, p, q \, \varepsilon \, N . \supset . \dfrac{p}{q} = \dfrac{mp}{mq} .$ (Euclid, VII, 17)

7. $\;\; p, q \, \varepsilon \, N . \, m \, \varepsilon \, N . \, m \, D \, p . \, m \, D \, q : \supset . \dfrac{p}{q} = \dfrac{p/m}{q/m} .$

8. $\;\; p, q, p', q' \, \varepsilon \, N . \, p \, \pi \, q . \, p' \, \pi \, q' . \dfrac{p}{q} = \dfrac{p'}{q'} : \supset : p = p' . \, q = q' .$

9. $\;\; p, q, p', q' \, \varepsilon \, N . \, p' \, \pi \, q' . \dfrac{p}{q} = \dfrac{p'}{q'} : \supset : p/p' = q/q' = M \, 3 \, D(p, q).$

10. $\;\; p, q, p', q' \, \varepsilon \, N . \dfrac{p}{q} = \dfrac{p'}{q'} . \, p \, \pi \, q . \, q' < q : = \Lambda.$ (Euclid, VII, 21)

11. $\;\; p, q, p', q' \, \varepsilon \, N : \supset : \dfrac{p}{q} = \dfrac{p'}{q'} . = . \dfrac{p}{p'} = \dfrac{q}{q'} . = . \dfrac{q}{p} = \dfrac{q'}{p'} .$
  (Euclid, VII, 13)

12. $\;\; p, q \, \varepsilon \, N . \supset :: [m \, \varepsilon] : m \, \varepsilon \, N . \, m \dfrac{p}{q} \, \varepsilon \, N \,\therefore\, - = \Lambda.$

12'. $\;\; a \, \varepsilon \, R . \supset :: [m \, \varepsilon] : m \, \varepsilon \, N . \, ma \, \varepsilon \, N \,\therefore\, - = \Lambda.$

13. $\;\; p, q, p', q' \, \varepsilon \, N . \supset :: [(r, s, t) \varepsilon] : r, s, t \, \varepsilon \, N . \dfrac{p}{q} = \dfrac{r}{t} . \dfrac{p'}{q'} = \dfrac{s}{t} \,\therefore\, - = \Lambda.$

13'. $a, b \ \varepsilon \ \mathrm{R} \ . \ \supset \ :: \ [(r, s, t)\varepsilon] : r, s, t \ \varepsilon \ \mathrm{N} \ . \ a = \dfrac{r}{t} \ . \ b = \dfrac{s}{t} \ \therefore \ - = \Lambda.$

14. $a, b, c \ \varepsilon \ \mathrm{R} \ . \ \supset \ :: \ [(m, n, p, q)\varepsilon] : m, n, p, q \ \varepsilon \ \mathrm{N} \ .$

$a = \dfrac{m}{q} \ . \ b = \dfrac{n}{q} \ . \ c = \dfrac{p}{q} \ \therefore \ - = \Lambda.$

15. $p, q, r \ \varepsilon \ \mathrm{N} \ . \ a = \dfrac{p}{r} \ . \ b = \dfrac{q}{r} : \supset : a = b \ . \ = \ . \ p = q.$

16. $m \ \varepsilon \ \mathrm{N} \ . \ a, b \ \varepsilon \ \mathrm{R} \ . \ a = b \ . \ mc \ \varepsilon \ \mathrm{N} : \supset \ . \ mb \ \varepsilon \ \mathrm{N}.$

17. $a, b, c \ \varepsilon \ \mathrm{R} \ . \ \supset \ \therefore \ a = a.$
  $\supset \ \therefore \ a = b \ . \ = \ . \ b = a.$
  $\supset \ \therefore \ a = b \ . \ b = c : \supset \ . \ a = c.$

18. $\mathrm{N} \supset \mathrm{R}.$

Definitions

19. $a, b \ \varepsilon \ \mathrm{R} \ . \ \supset \ :: \ a < b \ . \ = \ \therefore \ x \ \varepsilon \ \mathrm{N} \ . \ xa, xb \ \varepsilon \ \mathrm{N} : \supset_x \ . \ xa < xb.$

20. $a, b \ \varepsilon \ \mathrm{R} \ . \ \supset \ : b > a \ . \ = \ . \ a < b.$

Theorems

21. $p, q, r \ \varepsilon \ \mathrm{N} \ . \ a = \dfrac{p}{r} \ . \ b = \dfrac{q}{r} : \supset : a < b \ . \ = \ . \ p < q.$

22. $p, q, p', q' \ \varepsilon \ \mathrm{N} \ . \ \supset : \dfrac{p}{q} < \dfrac{p'}{q'} \ . \ = \ . \ pq' < p'q.$

23. $p, q, r \ \varepsilon \ \mathrm{N} \ . \ a = \dfrac{r}{p} \ . \ b = \dfrac{r}{q} : \supset : a < b \ . \ = \ . \ p > q.$

24. $p, q, p', q' \ \varepsilon \ \mathrm{N} \ . \ \dfrac{p}{q} < \dfrac{p'}{q'} : \supset \ . \ \dfrac{p}{q} < \dfrac{p + p'}{q + q'} < \dfrac{p'}{q'} \ .$

25. $a \varepsilon \ \mathrm{R} \ . \ \supset \ \therefore \ \mathrm{R} \ . \ \mathfrak{z} > a : - = \Lambda.$

26. $a \varepsilon \ \mathrm{R} \ . \ \supset \ \therefore \ \mathrm{R} \ . \ \mathfrak{z} < a : - = \Lambda.$

27. $a, b \varepsilon \ \mathrm{R} \ . \ a < b : \supset \ \therefore \ \mathrm{R} \ . \ \mathfrak{z} > a \ . \ \mathfrak{z} < b : - = \Lambda.$

28. $a, b \varepsilon \ \mathrm{R} : \supset \ \therefore \ a < b \ . \ a = b : \ = \Lambda.$
  $\supset \ \therefore \ a > b \ . \ a = b : \ = \Lambda.$
  $\supset \ \therefore \ a < b \ . \ a > b : \ = \Lambda.$
  $\supset \ \therefore \ a - \ < b \ . \ a - = b \ . \ a - \ > b : \ = \Lambda.$

29. $a, b, c \varepsilon \ \mathrm{R} : \supset \ \therefore \ a < \ \cup \ = b \ . \ b < c : \supset : a < c.$
  $\supset \ \therefore \ a < b \ . \ b < \ \cup = c : \supset : a < c.$

Definitions

30. $a, b \, \varepsilon \, R \, . \, \supset \, . \, a + b = [c\varepsilon](c \, \varepsilon \, R \, \therefore \, x \, \varepsilon \, N \, . \, xa, xb, xc \, \varepsilon \, N : \supset_x \, .$
$xa + xb = xc).$
31. $a, b \, \varepsilon \, R \, . \, \supset \, . \, b - a = [x\varepsilon](x \, \varepsilon \, R \, . \, a + x = b).$
32. $a, b \, \varepsilon \, R \, . \, \supset \, . \, ab = [c\varepsilon](c \, \varepsilon \, R \, \therefore \, x \, \varepsilon \, N \, . \, xa, (xa)b, xc \, \varepsilon \, N : \supset_x \, .$
$(xa)b = xc).$
33. $a, b \, \varepsilon \, R \, . \, \supset \, . \, b/a = [x\varepsilon](x \, \varepsilon \, R \, . \, ax = b).$

Theorems

34. $p, q, r \, \varepsilon \, N \, . \, \supset \, . \, \dfrac{p}{r} + \dfrac{q}{r} = \dfrac{p + q}{r} \, .$

35. $a, b \, \varepsilon \, R \, . \, \supset \, . \, a + b \, \varepsilon \, R.$

36. $p, q, r \, \varepsilon \, N \, . \, p < q : \supset \, . \, \dfrac{q}{r} - \dfrac{p}{r} = \dfrac{q - p}{r} \, .$

37. $a, b \, \varepsilon \, R \, . \, a < b : \supset \, . \, b - a \, \varepsilon \, R.$

38. $p, q, p', q' \, \varepsilon \, N \, . \, \supset \, . \, \dfrac{p}{q}\dfrac{p'}{q'} = \dfrac{pp'}{qq'} \, .$

39. $a, b \, \varepsilon \, R \, . \, \supset \, . \, ab \, \varepsilon \, R.$

40. $p, q, p', q' \, \varepsilon \, N \, . \, \supset \, . \, \dfrac{p}{q} \bigg/ \dfrac{p'}{q'} = \dfrac{pq'}{p'q} \, .$

41. $a, b \, \varepsilon \, R \, . \, \supset \, . \, b/a \, \varepsilon \, R.$

42. $p, q \, \varepsilon \, N \, . \, \supset \, . \, \dfrac{p}{q} = p/q.$

9 THE SYSTEM OF RATIONALS. IRRATIONALS

Explanation

If $a \, \varepsilon \, KR$, the sign Ta is read *upper boundary*, or *upper limit of the class a*. We shall define only a few relations and operations on this new entity.

Definitions

1. $a \, \varepsilon \, KR \, . \, x \, \varepsilon \, R : \supset :: x < Ta \, . \, = \, \therefore \, a \, . \, \ni > x : - \, = \Lambda.$
2. $a \, \varepsilon \, KR \, . \, x \, \varepsilon \, R : \supset ::: x = Ta \, . \, = \, ::: a \, . \, \ni > x : = \Lambda :: u \, \varepsilon \, R \, . \, u$
$< x : \supset_u \, \therefore \, a \, . \, \ni > u : - \, = \Lambda.$
3. $a \, \varepsilon \, KR \, . \, x \, \varepsilon \, R : \supset \, \therefore \, x > Ta \, . \, = \, : x - < Ta \, . \, x - \, = Ta.$

Theorem

4. $x \varepsilon R . \supset :: x = \therefore T : R . 3 < x.$

Explanation

The sign Q is read *quantity*, and indicates the positive real numbers, rational or irrational, with the exception of 0 and ∞.

Definitions

5. $Q = [x\varepsilon](a \varepsilon KR : a - = \Lambda : R 3 > Ta . - = \Lambda :$
   $Ta = x \therefore - = \Lambda).$
6. $a, b \varepsilon Q . \supset :: a = b . = \therefore R . 3 < a : = : R . 3 < b.$
7. $a, b \varepsilon Q . \supset :: a < b . = \therefore R . 3 > a . 3 < b : - = \Lambda.$
8. $a, b \varepsilon Q . \supset : b > a . = . a < b.$

Theorems

9. $a \varepsilon Q . \supset \therefore R . 3 < a : - = \Lambda.$
10. $a \varepsilon Q . \supset \therefore R . 3 > a : - = \Lambda.$
11. $R \supset Q.$

The propositions obtained from P 17, 28, 29 in §8 by reading Q for R also hold.

Definitions

12. $a, b \varepsilon Q . \supset . a + b = T[z\varepsilon]([(x, y)\varepsilon] : x, y \varepsilon R . x < a . y < b .$
    $x + y = z \therefore - = \Lambda).$
13. $a, b \varepsilon Q . \supset . ab = T[z\varepsilon]([(x, y)\varepsilon] : x, y \varepsilon R . x < a . y < b .$
    $xy = z \therefore - = \Lambda).$

In order for these definitions to have meaning, it must be proved that propositions 12 and 13 hold, if $a, b \varepsilon R$.

Subtraction and division could be defined as the inverse operations to addition and multiplication, and their properties could be proved.

10 SYSTEMS OF QUANTITIES

Explanations

If $a \varepsilon KQ$, the signs Ia, Ea, La are read: *interior, exterior, limit of the class a.*

Definitions

1. $a \, \varepsilon \, KQ . \supset . \, Ia = Q[x\varepsilon]([(u, v)\varepsilon] :: u, v \, \varepsilon \, Q \, \therefore \, u < x < v \, \therefore$
   $3 > u . 3 < v : \supset : a ::: - = \Lambda).$
2. $a \, \varepsilon \, KQ . \supset . \, Ea = I(-a).$
3. $a \, \varepsilon \, KQ . \supset . \, La = (-Ia)(-Ea).$

Theorems

4. $a \, \varepsilon \, KQ . x, u, v \, \varepsilon \, Q . u < x < v . (3 > u . 3 < v : \supset a) : \supset . x \, \varepsilon \, Ia.$
5. $a \, \varepsilon \, KQ . x \, \varepsilon \, Ia : \supset : [(u, v)\varepsilon](u, v \, \varepsilon \, Q \, \therefore \, u < x < v \, \therefore \, 3 > u .$
   $3 < v : \supset : a) - = \Lambda.$

PROOF $P1 = (P4)(P5).$

6. $a \, \varepsilon \, KQ . u, v \, \varepsilon \, Q . (3 > u . 3 < v : \supset a) \, \therefore \, \supset \, \therefore \, 3 > u . 3 < v : \supset \, Ia.$

PROOF $P6 = P4.$

7. $a \, \varepsilon \, KQ . \supset . \, Ia \supset a.$
8. $a \, \varepsilon \, KQ . \supset . \, IIa = Ia.$

PROOF

Hp. $(Ia)[a] \, P7 : \supset . \, IIa \supset Ia.$ (1)

Hp. $x, u, v \, \varepsilon \, Q . u < x < v . (3 > u . 3 < v : \supset a) . P6 : \supset :$
$u, v \, \varepsilon \, Q . u < x < v . (3 > u . 3 < v : \supset Ia).$ (2)

Hp. $x \, \varepsilon \, Ia . (2) : \supset : x \, \varepsilon \, IIa.$ (3)

Hp. $(3) : \supset : Ia \supset IIa.$ (4)

Hp. $(1) . (4) : \supset : Ts.$ (Theor.)

9. $a, b \, \varepsilon \, KQ . a \supset b : \supset . \, Ia \supset Ib.$

PROOF

Hp. $x, u, v \, \varepsilon \, Q . u < x < v . (3 > u . 3 < v : \supset a) : \supset \, \therefore$
$3 > u . 3 < v : \supset b.$ (1)

Hp. $x \, \varepsilon \, Ia : \supset : x \, \varepsilon \, Ib.$ (Theor.)

10. $a, b \, \varepsilon \, KQ : \supset : I (ab) \supset Ia.$

PROOF $(ab, a)[a, b] \, P9 . = . P10.$

11. $a, b \, \varepsilon \, KQ . \supset . \, I (ab) \supset (Ia)(Ib).$

PROOF P11 = : P10 . $\cap$. $(b, a)[a, b]$ P10.

12. $a, b \, \varepsilon \, KQ . \supset . \, Ia \supset I \, (a \cup b)$.

13. $a, b \, \varepsilon \, KQ . \supset . \, Ia \cup Ib \supset I \, (a \cup b)$.

14. $a, b \, \varepsilon \, KQ . \supset . \, I \, (ab) = (Ia)(Ib)$.

PROOF

Hp. P11 : $\supset$. I $(ab) \supset (Ia)(Ib)$.   (1)

Hp. $x \, \varepsilon \, Q . u, v \, \varepsilon \, Q . u < x < v . (\mathfrak{z} > u . \mathfrak{z} < v : \supset a) . u', v' \, \varepsilon \, Q$
    $. u' < x < v' . (\mathfrak{z} > u' . \mathfrak{z} < v' : \supset b) . u'' = M \, (u \cup u')$,
    $v'' = W \, (v, v') : \supset : u'', v'' \, \varepsilon \, Q . u'' < x < v'' .$
    $(\mathfrak{z} > u'' . \mathfrak{z} < v'' : \supset : ab)$.   (2)

Hp. $x \, \varepsilon \, Ia . x \, \varepsilon \, Ib . (2) : \supset . x \, \varepsilon \, I \, (ab)$.   (3)

Hp. (3) : $\supset$: $(Ia)(Ib) \supset I \, (ab)$.   (4)

Hp. (1) . (4) : $\supset$. Ts.

15. $a \, \varepsilon \, KQ . \supset . \, Ea \supset - a$.

PROOF P15 = $(-a)[a]$ P7.

16. $a \, \varepsilon \, KQ . \supset \therefore \, Ia . Ea : = \Lambda$.

PROOF Hp. P7 . P15 : $\supset \therefore \, Ia . Ea : \supset : a - a : = \Lambda$.

17. $a \, \varepsilon \, KQ . \supset . \, I \, Ea = Ea$.

PROOF P17 = $(-a)[a]$ P8.

18. $a, b \, \varepsilon \, KQ . b \supset a : \supset . \, Ea \supset Eb$.

PROOF P18 = $(-a, -b)[a, b]$ P9.

19. $a, b \, \varepsilon \, KQ . \supset : Ea \cup Eb . \supset E \, (ab)$.

20. $a, b \, \varepsilon \, KQ . \supset . \, E \, (a \cup b) = (Ea)(Eb)$.

PROOF P20 = $(-a, -b)[a, b]$ P14.

21. $a \, \varepsilon \, KQ . \supset . \, L \, (-a) = La$.

22. $a \, \varepsilon \, KQ . \supset \therefore \, Ia . La : = \Lambda$.
                       $\supset \therefore \, Ea . La : = \Lambda$.
                       $\supset \therefore \, -Ia . -Ea . -La : = \Lambda$.

PROOF P22 = P3.

23. $a \, \varepsilon \, KQ . \supset : a \supset . \, Ia \cup La$.

24. $a \, \varepsilon \, KQ . \supset . \, I \, (a \, La) = \Lambda$.

PROOF Hp. P14 . P7 . P22 : ⊃ : I $(a$ L$a)$. = . Ia ILa . ⊃ . IaLa . = . Λ.

25. a, b ε KQ . a ⊃ b : ⊃ : La . ⊃ . Ib ∪ Lb.

PROOF Hp. P18 : ⊃ : Eb ⊃ Ea : ⊃ : Ia ∪ La . ⊃ . Ib ∪ Lb : ⊃ . Ts.

26. a, b ε KQ . ⊃ : L (ab) ⊃ . IaLb ∪ IbLa ∪ LaLb.

PROOF

    Hp . ⊃ : ab ⊃ a . ab ⊃ b . P25 : ⊃ : L (ab) ⊃ Ia ∪ La . L (ab) ⊃ Ib
    ∪ Lb : ⊃ : L (ab) ⊃ (Ia ∪ La)(Ib ∪ Lb) . L (ab)(Ia)(Ib) =
    L (ab) I (ab) = Λ : ⊃ : Ts.

26'. a, b ε KQ . ⊃ . L (ab) ⊃ La ∪ Lb.

27. a, b ε KQ . ⊃ : L $(a ∪ b)$ ⊃ LaEb ∪ LbEa ∪ LaLb.

PROOF P27 = $(-a, -b)[a, b]$ P26.

27'. a, b ε KQ . ⊃ : L $(a ∪ b)$ ⊃ La ∪ Lb.

28. a ε KQ . ⊃ . LIa ⊃ La.

PROOF

    Hp. P7 : ⊃ : Ia ⊃ a . P25 : ⊃ : LIa ⊃ Ia ∪ La.   (1)
    Hp. P8 . P22 : ⊃ . LIaIa = LIaIIa = Λ.   (2)
    (1)(2) . ⊃ . Theor.

28'. a ε KQ . ⊃ . LEa ⊃ La.

29. a ε KQ . ⊃ . LLa ⊃ LIa ∪ LEa.

PROOF Hp. ⊃ : LLa = L $(I$a$ ∪ Ea)$. P27' : ⊃ . Ts.

29'. a ε KQ . ⊃ . LLa ⊃ La.

PROOF P29 . P28 . P28' : ⊃ . Theor.

30. a ε KQ . ⊃ . La = ILa ∪ LLa.

PROOF

    Hp. P23 : ⊃ . La ⊃ ILa ∪ LLa.   (1)
    Hp. P7 : ⊃ . ILa ⊃ La.   (2)
    Hp. P29' : ⊃ . LLa ⊃ La.   (3)
    (1)(2)(3) . ⊃ . Theor.

31. a ε KQ . ⊃ . LILa ⊃ LLa.

PROOF P31 = $(L$a$)[a]$ P28.

32. $a \varepsilon \mathrm{KQ} . \supset . \mathrm{ILL}a = \Lambda.$

PROOF Hp. P29' $: \supset : \mathrm{LL}a = \mathrm{La}\mathrm{LL}a . (\mathrm{L}a)[a]$ P24 $: \supset$ Ts.

33. $a \varepsilon \mathrm{KQ} . \supset : \mathrm{ILIL}a = \Lambda.$

PROOF P31 . P32 $: \supset$. P33.

34. $a \varepsilon \mathrm{KQ} . \supset . \mathrm{LLL}a = \mathrm{LL}a.$

PROOF $(\mathrm{L}a)[a]$ P30 . P32 $: \supset$ Theor.

35. $a, b \varepsilon \mathrm{KQ} . \supset . \mathrm{Ia}\mathrm{L}b \supset \mathrm{L}(ab).$

PROOF

    Hp. P14 $: \supset . \mathrm{Ia}\mathrm{L}b\mathrm{I}(ab) = \mathrm{Ia}\mathrm{I}b\mathrm{L}b = \Lambda.$ (1)

    Hp. P2 . P14 $: \supset . \mathrm{Ia}\mathrm{L}b\mathrm{E}(ab) = \mathrm{Ia}\mathrm{L}b\mathrm{I}(-a \cup -b) =$

    $\mathrm{I}(a - b)\mathrm{L}b = \mathrm{Ia}\mathrm{E}b\mathrm{L}b = \Lambda.$ (2)

    (1)(2) $\supset$ Theor.

36. $a, b \varepsilon \mathrm{KQ} . \supset . \mathrm{Ia}\mathrm{L}b \cup \mathrm{I}b\mathrm{La} \supset \mathrm{L}ab.$ (See P26)

PROOF P36 $= :$ P35 . $(b, a)[a, b]$ P35.

37. $a, b \varepsilon \mathrm{KQ} . \supset . \mathrm{Ea}\mathrm{L}b \cup \mathrm{E}b\mathrm{La} \supset \mathrm{L}(a \cup b).$ (See P27)

PROOF P37 $= (-a, -b)[a, b]$ P36.

38. $a, b \varepsilon \mathrm{KQ} . \supset . \mathrm{I}(a \cup b) \supset \mathrm{Ia} \cup \mathrm{I}b \cup \mathrm{La}\mathrm{L}b.$ (See P13)

PROOF

    Hp. $\supset . \mathrm{I}(a \cup b) \supset (\mathrm{Ia} \cup \mathrm{La} \cup \mathrm{Ea})(\mathrm{I}b \cup \mathrm{L}b \cup \mathrm{E}b).$ (1)

    Hp. P20 . P16 $: \supset . \mathrm{I}(a \cup b)\mathrm{Ea}\mathrm{E}b = \mathrm{I}(a \cup b)\mathrm{E}(a \cup b) = \Lambda.$ (2)

    Hp. P37 $: \supset : \mathrm{I}(a \cup b)(\mathrm{Ea}\mathrm{L}b \cup \mathrm{E}b\mathrm{La}) . \supset .$

    $\mathrm{I}(a \cup b)\mathrm{L}(a \cup b) . = \Lambda.$ (3)

    (1)(2)(3) . $\supset$. Theor.

38'. $a, b \varepsilon \mathrm{KQ} . \supset . \mathrm{E}(ab) \supset \mathrm{Ea} \cup \mathrm{E}b \cup \mathrm{La}\mathrm{L}b.$ (See P19)

39. $a \varepsilon \mathrm{KQ} . \supset . \mathrm{ILa}\mathrm{LIa} = \Lambda.$

PROOF Hp. P36 $: \supset : \mathrm{ILa}\mathrm{LIa} \supset \mathrm{L}(\mathrm{LaIa}) = \Lambda.$

40. $a \varepsilon \mathrm{KQ} . \supset . \mathrm{LIa} \supset \mathrm{LL}a.$

PROOF Hp. P28 . P30 . P39 $: \supset$ Theor.

40'. $a \varepsilon \mathrm{KQ} . \supset . \mathrm{LEa} \supset \mathrm{LL}a.$

41. $a \, \varepsilon \, KQ \, . \, \supset \, . \, LLa = LIa \cup LEa.$

PROOF P29 . P40 . P40′ : $\supset$. Theor.

42. $a \, \varepsilon \, KQ \, . \, \supset \, . \, ILIa = \Lambda.$
$\supset . \, ILEa = \Lambda.$
$\supset . \, LLIa = LIa.$
$\supset . \, LLEa = LEa.$

43. $a, b \, \varepsilon \, KQ \, . \, \supset \, . \, I \, (Ia \cup Ib) = Ia \cup Ib.$

PROOF

Hp. P7 : $\supset . \, I \, (Ia \cup Ib) \supset Ia \cup Ib.$ (1)

Hp. P8 . P13 : $\supset$: $Ia \cup Ib . = . \, IIa \cup IIb . \supset . \, I \, (Ia \cup Ib).$ (2)

(1)(2) $\supset$ Theor.

44. $a, b \, \varepsilon \, KQ \, . \, \supset \, . \, I \, (LLa \cup LLb) = \Lambda.$

PROOF

Hp. P38 . P32 . P34 : $\supset . \, I \, (LLa \cup LLb) \supset LLaLLb \supset LLa.$ (1)

Hp. (1) . P8 : $\supset . \, I \, (LLa \cup LLb) \supset ILLa = \Lambda.$

45. $a \, \varepsilon \, KQ \, . \, \supset \, . \, I \, (Ia \cup Ea) = Ia \cup Ea.$

PROOF P8 . P17 . $(-a)[b]$ P43 : $\supset$. Theor.

45′. $a \, \varepsilon \, KQ \, . \, \supset \, . \, ELa = Ia \cup Ea.$

46. $a \, \varepsilon \, KQ \, . \, \supset \, . \, EIa = -(Ia \cup LIa).$

46′. $a \, \varepsilon \, KQ \, . \, \supset \, . \, EEa = -(Ea \cup LEa).$

VIII
An approximation formula for the perimeter of the ellipse (1889)*

On 26 November 1889 Peano wrote to Felice Casorati regarding a note that Casorati had offered to present to the Accademia dei Lincei, and he added:

And seeing that you are so very kind, I'll ask another bit of advice. Mr Boussinesq, in the *Comptes rendus* (Académie des Sciences, Paris) of this year, on page 695, gives as new the formula for approximating the perimeter of an ellipse (with semiaxes a and b)

$$\pi \left(3 \frac{a + b}{2} - \sqrt{ab} \right),$$

which is identical with that given by me, two years ago, in my *Applicazioni geometriche*, page 233, in the form

$$\pi(a + b) + \tfrac{1}{2}\pi(\sqrt{a} - \sqrt{b})^2.$$

Does it seem to you, Professor, that the question is important enough for me to vindicate my priority – or is it not worth the effort?

It is not known what, if anything, was Casorati's reply, but Peano lost no time and on 25 December could report to Casorati: 'My observations on the approximate rectification of the ellipse have already been presented to the Academy of Sciences of Paris.' Casorati answered the following day, saying that he was pleased that Peano's note had been presented to the Academy and added: 'Boussinesq was very pleased with that formula, but he is a good man and can find your vindication only just.'

The following selection is the note which Hermite presented to the Academy of Sciences.

The approximation formula for the length E of an ellipse

$$\pi \left(3 \frac{a + b}{2} - \sqrt{ab} \right)$$

---

* 'Sur une formule d'approximation pour la rectification de l'ellipse,' *Comptes rendus des séances de l'Académie des sciences de Paris*, 109 (1889), 960–1 [22].

given by Mr Bussinesq in the *Comptes rendus* (108 (1889), 695) had already been published by me in my *Applicazioni geometriche del calcolo infinitesimale* (Turin, 1887, p. 233) in the form

(1) $E < \pi(a + b) + \frac{1}{2}\pi(\sqrt{a} - \sqrt{b})^2.$

I found this formula by developing the radical in the expression

$$E = 4 \int_0^{\frac{1}{2}\pi} \sqrt{a^2\cos^2 t + b^2\sin^2 t} \, dt$$

in a continued fraction

$$\alpha + \cfrac{\beta}{2\alpha + \cfrac{\beta}{2\alpha + \ldots}},$$

where

$$\alpha = a \cos^2 t + b \sin^2 t, \qquad \beta = (a - b)^2 \sin^2 t \cos^2 t.$$

If we integrate the first convergent, we get

$$E > \pi(a + b);$$

the second gives formula (1). From the third convergent of the continued fraction we deduce the new formula

$$E > \frac{2\pi}{9}\left(19s - 4\frac{2s^2 + 3p^2}{\sqrt{s^2 + 3p^2}}\right),$$

where $2s = a + b$, $p^2 = ab$. If we take the second member as an approximate value of E, the error is less than

$$\frac{5\pi(a - b)^6}{16384b^5}.$$

IX

On the definition of the area
of a surface (1890, 1903)*

In 1882, while substituting for Angelo Genocchi in the calculus course at the University of Turin, Peano discovered an error in J.A. Serret's commonly accepted definition of the area of a curved surface. In the following selection Peano gives his analysis of the difficulty involved in defining the area of a curved surface and gives his solution (and asserts his priority of publication, omitting the fact that H.A. Schwarz discovered the error first). This article influenced Henri Lebesque in the writing of his Ph.D. dissertation, 'Intégrale, Longueur, Aire' (Ann. mat. pura appl., (3) 7 (1902), 231–359).

We follow this with an extract from the Formulaire mathématique *of 1903, which contains the counterexample (to Serret's definition of the area of a curved surface) referred to in the preceding article.*

1 ON THE DEFINITION OF THE AREA OF A SURFACE

The purpose of the present note is to examine several definitions of the area of a region of a (non-planar) surface, along with some related questions.

The Greek geometers, reasoning about the length of lines and areas of surfaces (spheres, cylinders, etc.), started out from postulates instead of definitions. The difference, however, is purely formal. The postulates stated by Archimedes[1] are exactly equivalent to the following definitions:

1. The length of a convex arc in a plane is the common value of the least upper bound (l.u.b.) of the lengths of the inscribed polygonal curves and the greatest lower bound (g.l.b.) of the circumscribed polygonal curves.

2. The area of a convex surface is the common value of the l.u.b. of the areas of the inscribed polyhedral surfaces and the g.l.b. of the circumscribed polyhedral surfaces. He proves the coincidence of these two limits for the curves and surfaces studied, but this could also be proved in general.

* 'Sulla definizione dell'area d'una superficie,' *Atti Accad. naz. Lincei, Rend., Cl. sci. fis. mat. nat.*, (4) 6-I (1890), 54–7 [23], and extract from *Formulaire mathématique*, IV (Turin: Bocca Frères, 1903), pp. 300–1 [125].
1 'On the Sphere and Cylinder,' Book I, λαμβανόμενα [= things assumed].

Now, the first definition does not hold for non-planar curves. It can be made applicable in every case by omitting the circumscribed curves:

3. The length of a curve is the l.u.b. of the lengths of its inscribed polygonal curves.[2]

The second definition, that of area, is not applicable for general surfaces and it appears difficult to make it applicable in every case.

The procedures for determining the length of an arc and the area of a surface followed by various mathematicians up to the beginning of this century were never very exact.[3] Only relatively recently has it been usual for calculus texts to give definitions of the length of an arc and the area of a surface. Now, although the definition of the first may not present difficulties, the definition of the second always leaves something to be desired. The definition given by Serret (and copied by many authors) in which one considers the limit toward which the inscribed polyhedral surfaces tend is not valid, because such a polyhedral surface may tend, depending on the way it varies, toward any limit which is larger than that quantity which everyone says is the area of the surface.[4]

The late Professor Harnack, in his version of Serret's text,[5] adds the condition that the faces of the polyhedral surface are to be triangles whose angles do not approach zero. But even with this condition the definition is not satisfactory, since the same objection may still be raised.

Mr Hermite[6] says, 'We shall therefore not use the polyhedral surface, which is the analogue of the inscribed polygon of an arc of a curve ...,' and

2 This definition is simpler than the usual one, the concept of 'l.u.b. of a set of quantities' being simpler than that of 'limit toward which a variable quantity tends.' From it we immediately deduce that every arc has a finite or infinite length. I earlier used such a definition in my *Applicazioni geometriche del calcolo differenziale*, p. 161. Mr Jordan, in his *Cours d'analyse*, vol. III (1887), p. 594, proves the coincidence of the two definitions in the most common cases.

3 For example, in Lagrange, *Théorie des fonctions analytiques* (Paris, 1813), p. 300, the result is obtained by means of an inexact assertion.

4 This observation was published for the first time in the lessons given by me at the University of Turin in the year 1881–2 and lithographed by the students. It is found on p. 143 of the lesson of 22 May 1882. The same observation was also made by Mr Schwarz, who communicated it to Mr Hermite, who then published it in his *Cours professé à la faculté des sciences, pendant le second semestre 1882* (second printing, p. 35) some time after my publication. Serret's principal error consists in holding that the plane passing through three points of a surface has as its limit the plane tangent to that surface, a proposition which is obviously false.

5 *Lehrbuch der Differential- und Integralrechnung*, vol. II (1885), p. 295. From the condition imposed by Harnack it follows that the planes of the faces tend toward the tangent plane. The defect lies in this, that if $z = f(x, y)$ is the equation of the surface and $f(x, y)$ is a single-valued function, it does not follow that every inscribed polyhedral surface cannot be intersected in more than one point by a parallel to the z-axis.

6 *Ibid.*, third edition (1887), p. 36.

defines the area as the limit of a system of non-contiguous polygons which are tangent to the surface. This definition, otherwise quite rigorous, leaves something to be desired in that the reference axes enter into it explicitly.

Rigour and the analogy between the definitions relating to arc and area may be both preserved by making use not only of the concept of a straight-line segment considered as having length and direction (vector), but also of the dual concept of a planar region considered as having size and orientation. These entities were introduced into geometry through the works of Chelini, Möbius, Bellavitis, Grassmann, and Hamilton. A planar region so considered, or rather its boundary, may be said to be a bivector, it being the product, in the sense of Grassmann, of two vectors.[7]

We then have the proposition:

4. Given a closed (non-planar) curve l, a closed planar curve or bivector l' can always be determined such that, if the two curves l and l' are projected onto an arbitrary plane, with parallel rays of an arbitrary direction, then the limiting areas of their projections are always equal.

This proposition is an immediate consequence of the sum, or composition, of bivectors when the curve l is polygonal. The usual passage to the limit permits its proof when the curve l is described by a point having everywhere a finite derivative, and in other cases also. In considering areas one must always take the signs into account.

By virtue of proposition 4 we may say that every closed curve, whether planar or not, is a bivector. Two bivectors l and l' which satisfy the conditions of proposition 4 are said to be equal. By size and orientation of a non-planar bivector l we mean the size and orientation of the equal planar bivector l'.[8]

It is clear that:

5. If the closed curve l is projected orthogonally onto a variable plane, then the maximum area of the projection of l equals the size of the bivector l, and this maximum is obtained when the plane onto which the curve is projected has the orientation of l.

If we understand by 'vector of an arc of a curve' the vector limited by the ends of the arc, that is, if we consider its chord as a vector, then definition 3 may also be stated:

6. The length of an arc of a curve is the l.u.b. of the sum of the lengths of the vectors of its parts.

---

7 I used the name bivector, corresponding to that of vector introduced by Hamilton, in my *Calcolo geometrico, secondo l'Ausdehnungslehre di H. Grassmann* (1888).

8 Proposition 4 is useful in many geometrical questions. For example, consider a helical spiral, the radii to its ends, and that part of the axis connecting them. This is a closed curve, and it can easily be seen that it is equal to the circumference of the base of the cylinder on which the helix rests. Projecting two curves onto planes by parallel rays in this way, we may deduce the areas enclosed by various planar curves.

Analogously, if we understand by 'bivector of a region of a surface' the bivector formed by its boundary, then we may give the definition:

7. The area of a region of a surface is the l.u.b. of the sum of the sizes of the bivectors of its parts.[9]

Between the vector of an arc of a curve and the bivector of a region of a surface there is a complete analogy. Thus, to the proposition that, under certain conditions,

8. The direction of the vector of an infinitesimal arc of a curve is that of the tangent, and the ratio of its length to the length of the arc is unity, corresponds the proposition that, under analogous conditions,

9. The orientation of the bivector of an infinitesimal region of a surface is that of the tangent plane, and the ratio between its size and the area of that region is unity.

The proposition:

10. The first term in the development of the difference between an arc s and its chord, in ascending powers of s, is

$$s^3/24R^2,$$

where R is the radius of curvature, has the analogue:

11. The first term in the development in powers of ρ, of the difference between the area of a geodetic circle, traced on a surface of radius ρ, and the size of its bivector is

$$\frac{\pi\rho^4}{8}\left(\frac{1}{R_1{}^2} + \frac{1}{R_2{}^2}\right) = \frac{\pi\rho^4}{4}\,C,$$

where R_1 and R_2 are the principal radii of curvature and C is the curvature of the surface according to the definition of Professor Casorati.[10]

2 THE DEFINITION OF THE AREA OF A CURVED SURFACE

* $6 \cdot 0$ $u\varepsilon$ Cls'p. Volum $u = 0$. $\supset$.

Area $u = \lim\{\text{Volum}\{p \cap x3[d(x, u) < h]\}/(2h)|h, Q, 0\}$ Df

This is the definition of the area of a curved surface.

The area of a curved surface may not be defined as the limit of the areas of the inscribed polyhedral surfaces, for the faces of a polyhedron do not necessarily have as limits the planes tangent to the surface.

9 By replacing 'the size of a bivector' by its meaning, we transform this definition into that given by me in *Applicazioni geometriche*, p. 164.

10 'Mesure de la courbure des surfaces suivant l'idée commune,' *Acta mathematica*, vol. 14 (1890), Stockholm.

It is necessary to require that the planar angles of the faces do not become infinitely small.

For example, suppose:

$$o \, \varepsilon \, \mathrm{p} \,.\, i, j, k \, \varepsilon \, \mathrm{v} \,.\, i^2 = j^2 = k^2 = 1 \,.\, i \times j = j \times k = k \times i = 0 \,.$$
$$u = i/j \,.\, m, n \, \varepsilon \, \mathrm{N}_1 \,.\, r \, \varepsilon \, 0 \,...\, (m - 1) \,.\, s \, \varepsilon \, 0 \,...\, (n - 1).$$

Now consider the triangle whose vertices are

$$o + e^{2ru\pi/m}i + ks/n, \qquad o + e^{2(r+1)u\pi/m}i + ks/n,$$
$$o + e^{(2r+1)u\pi/m}i + k(s + 1)/n,$$

and the triangle whose vertices are

$$o + e^{2ru\pi/m}i + ks/n, \qquad o + e^{(2r-1)u\pi/m}i + k(s + 1)/n,$$
$$o + e^{(2r+1)u\pi/m}i + k(s + 1)/n.$$

As r and s vary, these triangles form a polyhedral surface inscribed in the cylinder having ok as axis, radius 1, and height 1. If the evolute of the cylindrical surface is taken, the vertices of the triangles form the adjacent figure ($m = 5$, $n = 4$.) The same figure is also obtained by taking the evolute of the polyhedral surface.

The area of the polyhedral surface is easily calculated to be

$$2m \sin(\pi/m) \, \sqrt{\{1 + 4n^2 \sin[\pi/(2m)]^4\}}$$

and its limit for $m = \infty, n = \infty$ is not determined. If one takes $n = m$, the limit is 2π, the area of the cylinder. If one takes $n = m^2$, then the limit is $2\pi \sqrt{(1 + \pi^4/4)}$. If $n = m^3$, then the limit is ∞.

The preceding calculation is taken from my lesson at the University of Turin of 22 May 1882, p. 143 of the lithograph of the course for 1881–2. See also: 'Sulla definizione dell'area d'una superficie,' *Atti Accad. naz. Lincei*, 19 January 1890 [i.e., the preceding article].

The same remark was made about the same time, and independently, by H.A. Schwarz (*Mathematische Abhandlungen*, Berlin, 2 [1890], p. 309).

In my note there is a definition which is analogous to the definition of arc and which is founded on the notion of bivector. We have chosen to give a more elementary one here.

Some authors, in considering S, project the region onto a plane, decompose the projected region into infinitely small elements ΔA, and define as the area of S the limit of $\sum \Delta A/\cos \gamma$, where γ is the angle formed by the plane of projection and the plane tangent to the surface at an arbitrary point of ΔA.

This definition is not homogeneous, since it explicitly contains the plane of projection. One ought to say: 'the area is the constant value of the limit considered, whatever the plane of projection.'

The definition given here was recognized as possible by Borchardt, *J. math. pures appl.*, 19 (1854), 369; *Werke*, p. 67: 'We know that the volume contained between two parallel surfaces reduces to the product of the area of one of the two surfaces by their distance, whenever this last becomes infinitely small. The inverse of this result shows that the area of a surface can be considered as the limit toward which converges the ratio whose numerator is the volume contained between the surface and its parallel, and whose denominator is the distance between the two surfaces.'

Without noticing this passage of Borchardt, this same definition was proposed by Minkowski, *Jber. Deutsch. math. Verein.*, 9 (1901), 115. (See also *R.d.M.*, 7 (1900–1), 109.) H. Minkowski also gave an analogous definition for the length of a class of points which coincides more or less with proposition P7 · 0 [of *Formulaire mathématique*, IV].

X

A space-filling curve (1890, 1908)*

The following selection contains what is probably, after the postulates for the natural numbers, Peano's best-known discovery. The 'curve which completely fills a planar region' was a spectacular counterexample to the commonly accepted notion that an arc of a curve given by continuous parametric functions could be enclosed in an arbitrarily small region. Indeed, here is a curve given by continuous parametric functions, x = x(t) and y = y(t), such that as t varies throughout the unit interval, the graph of the curve includes every point in the unit square.

We couple this selection with an excerpt from the Formulario *of 1908 in which Peano gives a graphical representation of one 'approach' to such a curve. Hilbert had, after the publication of Peano's original paper, given the first such graphical representation, but Peano was probably led to his discovery by just such a representation. He published his result without diagrams because, it would seem, he wanted no one to think that a false proof lurked in a forced interpretation of a diagram. His proof is purely analytic.*

Peano's curve is a mapping of the unit interval onto its Cartesian product. In the development of topology, this gave rise to the study of Peano *spaces. (A Peano space is a Hausdorff space which is an image of the unit interval under a continuous mapping.) It also raised the question: Which spaces can be mapped continuously onto their Cartesian product? It is of interest that Peano's example is unique, in the sense that: 'The only non-degenerate ordered continuum C which admits a mapping f : C → C × C onto its square C² = C × C is the real line segment I.'* (S. Mardešić, 'Mapping Ordered Continua onto Product Spaces,' *Glasnik mat.-fiz. astr. drustvo mat. fiz. hrvatske,* (2) 15 (1960), 85–9; p. 88, Theorem 4.)

1 ON A CURVE WHICH COMPLETELY FILLS A PLANAR REGION

In this note we determine two single-valued and continuous functions

---

* 'Sur une courbe, qui remplit toute une aire plane,' *Mathematische Annalen*, 36 (1890), 157–60 [24], and excerpt from *Formulario mathematico*, vol. 5 (Turin: Bocca, 1905–8), pp. 239, 240 [138].

x and y of a (real) variable t which, as t varies throughout the interval $(0, 1)$, take on all pairs of values such that $0 \leqslant x \leqslant 1, 0 \leqslant y \leqslant 1$. If, according to common usage, we consider the locus of points whose coordinates are continuous functions of a variable to be a continuous curve, then we have an arc of a curve which goes through every point of a square. Thus, being given an arc of a continuous curve, with no other hypothesis, it is not always possible to enclose it in an arbitrarily small region.

We shall use the number 3 as a base of numeration and refer to each of the numerals 0, 1, 2 as a digit. We now consider the infinite sequence of digits $a_1, a_2, a_3, ...$, which we write

$$T = 0 \,.\, a_1 a_2 a_3 \,...$$

(for the moment, T is merely a sequence of digits).

If a is a digit, we designate by $\mathbf{k}a$ the digit $2 - a$, the *complement* of a; i.e., we let

$$\mathbf{k}0 = 2, \qquad \mathbf{k}1 = 1, \qquad \mathbf{k}2 = 0.$$

If $b = \mathbf{k}a$, we deduce that $a = \mathbf{k}b$. We also have $\mathbf{k}a \equiv a \pmod 2$. We designate by $\mathbf{k}^n a$ the result of the operation $\mathbf{k}$ repeated n times on a. If n is even, we have $\mathbf{k}^n a = a$; if n is odd, $\mathbf{k}^n a = \mathbf{k}a$. If $m \equiv n \pmod 2$, we have $\mathbf{k}^m a = \mathbf{k}^n a$.

We let correspond to the sequence T the two sequences

$$X = 0 \,.\, b_1 b_2 b_3 \,..., \qquad Y = 0 \,.\, c_1 c_2 c_3 \,...,$$

where the digits b and c are given by the relations

$$b_1 = a_1, \quad c_1 = \mathbf{k}^{a_1} a_2, \quad b_2 = \mathbf{k}^{a_2} a_3, \quad c_2 = \mathbf{k}^{a_1 + a_3} a_4,$$
$$b_3 = \mathbf{k}^{a_2 + a_4} a_5, \,...,$$
$$b_n = \mathbf{k}^{a_2 + a_4 + ... + a_{2n-2}} a_{2n-1}, \quad c_n = \mathbf{k}^{a_1 + a_3 + ... + a_{2n-1}} a_{2n}.$$

Thus, b_n, the nth digit of X, is equal to a_{2n-1}, the nth digit of uneven rank in T, or to its complement, according as the sum $a_2 + ... + a_{2n-2}$ of digits of even rank, which precede it, is even or odd, and analogously for Y. We may thus write these relations in the form:

$$a_1 = b_1, \quad a_2 = \mathbf{k}^{b_1} c_1, \quad a_3 = \mathbf{k}^{c_1} b_2, \quad a_4 = \mathbf{k}^{b_1 + b_2} c_2, \,...,$$
$$a_{2n-1} = \mathbf{k}^{c_1 + c_2 + ... + c_{n-1}} b_n, \quad a_{2n} = \mathbf{k}^{b_1 + b_2 + ... + b_n} c_n.$$

If the sequence T is given, then X and Y are determined, and if X and Y are given, then T is determined.

We give the name *value* of the sequence T to the quantity (analogous to a decimal number having the same notation)

$$t = \text{val } T = \frac{a_1}{3} + \frac{a_2}{3^2} + \dots + \frac{a_n}{3^n} + \dots$$

To each sequence T corresponds a number t such that $0 \leqslant t \leqslant 1$. For the converse, the numbers t in the interval $(0, 1)$ divide into two classes:

(α) The numbers, different from 0 and 1, which give an integer when multiplied by a power of 3. They are represented by two sequences, the one

$$T = 0 . a_1 a_2 \dots a_{n-1} a_n 222 \dots,$$

where a_n is equal to 0 or 1, and the other

$$T' = 0 . a_1 a_2 \dots a_{n-1} a_n' 000 \dots,$$

where $a_n' = a_n + 1$.

(β) The other numbers. These are represented by only one sequence T.

Now, the correspondence established between T and (X, Y) is such that if T and T' are two sequences of different form, but val $T = $ val T', and if X, Y are the sequences corresponding to T, and X', Y' those corresponding to T', we have

$$\text{val } X = \text{val } X', \qquad \text{val } Y = \text{val } Y'.$$

Indeed, consider the sequence

$$T = 0 . a_1 a_2 \dots a_{2n-3} a_{2n-2} a_{2n-1} a_{2n} 2222 \dots,$$

where a_{2n-1} and a_{2n} are not both equal to 2. This sequence can represent every number of class α. Letting

$$X = 0 . b_1 b_2 \dots b_{n-1} b_n b_{n+1} \dots$$

we have

$$b_n = \mathbf{k}^{a_2 + \dots + a_{2n-2}} a_{2n-1}, \qquad b_{n+1} = b_{n+2} = \dots = \mathbf{k}^{a_2 + \dots + a_{2n-2} + a_{2n}} 2.$$

Letting T' be the other sequence whose value coincides with val T, we have

$$T' = 0 . a_1 a_2 \dots a_{2n-3} a_{2n-2} a_{2n-1}' a_{2n}' 0000 \dots$$

and

$$X' = 0 . b_1 \dots b_{n-1} b_n' b_{n+1}' \dots$$

The first $2n - 2$ digits of T' coincide with those of T; hence the first $n - 1$ digits of X' coincide also with those of X. The others are determined by the relations

$$b_n' = \mathbf{k}^{a_2 + \cdots + a_{2n-2}} a'_{2n-1},$$
$$b'_{n+1} = b'_{n+2} = \cdots = \mathbf{k}^{a_2 + \cdots + a_{2n-2} + a'_{2n}} 0.$$

We now distinguish two cases, according as $a_{2n} < 2$ or $a_{2n} = 2$. If a_{2n} has the value 0 or 1, we have

$$a'_{2n} = a_{2n} + 1, \qquad a'_{2n-1} = a_{2n-1}, \qquad b_n' = b_n,$$
$$a_2 + a_4 + \cdots + a_{2n-2} + a'_{2n} = a_2 + \cdots + a_{2n-2} + a_{2n} + 1,$$

whence

$$b'_{n+1} = b'_{n+2} = \cdots = b_{n+1} = b_{n+2} = \cdots = \mathbf{k}^{a_2 + \cdots + a_{2n}} 2.$$

In this case the two series X and X' coincide in form and value.

If $a_{2n} = 2$, we have $a_{2n-1} = 0$ or 1, $a'_{2n} = 0$, $a'_{2n-1} = a_{2n-1} + 1$, and on setting

$$s = a_2 + a_4 + \cdots + a_{2n-2}$$

we have

$$b_n = \mathbf{k}^s a_{2n-1}, \qquad b_{n+1} = b_{n+2} = \cdots = \mathbf{k}^s 2,$$
$$b_n' = \mathbf{k}^s a'_{2n-1}, \qquad b'_{n+1} = b'_{n+2} = \cdots = \mathbf{k}^s 0.$$

Now, since $a'_{2n-1} = a_{2n-1} + 1$, the two fractions $0 \, . \, a_{2n-1} 222...$ and $0 \, . \, a'_{2n-1} 000...$ have the same value. Operating on the digits with the same operation $\mathbf{k}^s$ we obtain the two fractions

$$0 \, . \, b_n b_{n+1} b_{n+2} \cdots \qquad \text{and} \qquad 0 \, . \, b_n' b'_{n+1} b'_{n+2} \cdots,$$

which also have the same value, as may easily be seen. Hence, the fractions X and X', although differing in form, have the same value.

Analogously we may show that val Y' = val Y.

Therefore, if we set x = val X, and y = val Y, we deduce that x and y are two single-valued functions of the variable t in the interval $(0, 1)$. Indeed, if t tends to t_0, the first $2n$ digits of the development of t finally coincide with those of the development of t_0 if t_0 is a β, or with those of one of the two developments of t_0 if t_0 is an α; and so the first n digits of the x and y corresponding to t will coincide with those of the x and y corresponding to t_0.

Finally, to each pair (x, y) such that $0 \leqslant x \leqslant 1, 0 \leqslant y \leqslant 1$ corresponds at least one pair of sequences (X, Y) which express that value; to (X, Y)

corresponds a T, and to this a t. Thus one may always determine t in such a fashion that the two functions x and y take on any arbitrarily given values in the interval $(0, 1)$.

We arrive at the same results if, in place of 3, we take any odd number whatever as our base of numeration. One could also take an even number as base, but then the correspondence between T and (X, Y) has to be less simple.

We may form an arc of a continuous curve which completely fills a cube. We make correspond to the fraction (to base 3)

$$T = 0 \,.\, a_1 a_2 a_3 a_4 \,...$$

the fractions

$$X = 0 \,.\, b_1 b_2 \,..., \qquad Y = 0 \,.\, c_1 c_2 \,..., \qquad Z = 0 \,.\, d_1 d_2 \,...,$$

where

$$b_1 = a_1, \qquad c_1 = \mathbf{k}^{b_1} a_2, \qquad d_1 = \mathbf{k}^{b_1 + c_1} a_3, \qquad b_2 = \mathbf{k}^{c_1 + d_1} a_4, \,...,$$

$$b_n = \mathbf{k}^{c_1 + ... + c_{n-1} + d_1 + ... + d_{n-1}} a_{3n-2},$$

$$c_n = \mathbf{k}^{d_1 + ... + d_{n-1} + b_1 + ... + b_n} a_{3n-1},$$

$$d_n = \mathbf{k}^{b_1 + ... + b_n + c_1 + ... + c_n} a_{3n}.$$

We then prove that $x = \text{val } X$, $y = \text{val } Y$, $z = \text{val } Z$ are single-valued and continuous functions of the variable $t = \text{val } T$; and that if t varies between 0 and 1, then x, y, z take on all triplets of values which satisfy the conditions $0 \leqslant x \leqslant 1, 0 \leqslant y \leqslant 1, 0 \leqslant z \leqslant 1$.

Mr Cantor (*Journal de Crelle*, 84 [1878], 242) has shown that a one-to-one correspondence (*unter gegenseitiger Eindeutigkeit*) between the points of a line and those of a surface can be established. But Mr Netto (*J. reine angew. Math.* 86 [1879], 263–8) and others have shown that such a correspondence is necessarily discontinuous. (See also G. Loria, 'La definizione dello spazio ad n dimensioni ... secondo le ricerche di G. Cantor,' *Giornale di matematiche*, 1877.) My note shows that one can establish single-valuedness and continuity from one side, i.e. to the points of a line can be made to correspond the points of a surface in such a fashion that the image of the line is the entire surface, and that the point on the surface is a continuous function of the point on the line. But this correspondence is not one-to-one, for to the points (x, y) of the square, if x and y are βs, then indeed there is only one corresponding t, but if x or y, or both, are αs, then the corresponding values of t are two or four in number.

It has been shown that an arc of a continuous planar curve may be enclosed in an arbitrarily small region:

1. If one of the functions, x for example, coincides with the independent variable t, in which case we have the theorem on the integrability of continuous functions.

2. If the two functions x and y are of limited variation (Jordan, *Cours d'analyse*, III [1887], p. 599). This is not true, however, as the preceding example shows, on the sole supposition of continuity of the functions x and y.

These x and y, the continuous functions of the variable t, are nowhere differentiable.

2 REMARKS ON A SPACE-FILLING CURVE

* 3. $n \, \varepsilon \, N_1 \, . \, \supset \, . \, \exists \, (Cxnfq) \, \text{cont} \, \cap f \, 3 \, (f\text{'q} = Cxn)$

There exists a complex of order n, or a point in n-dimensional space, which is a continuous function of a real variable, or of time, such that the trajectory of the moving point fills the whole space. That is, there exists a continuous curve which goes through every point of a plane, and there exists a curve which goes through every point of a space, etc. This result is of interest in the study of the foundations of geometry, for there does not exist a specific character which distinguishes a curve from a surface.

If we wish, as the variable t varies from 0 to 1, to have the point with coordinates x and y, functions of t, describe the whole square ($\Theta \colon \Theta$), we develop t as a decimal fraction, or to some base analogous to the decimal:

$$t = 0 \, . \, a_1 a_2 a_3 \, ...$$

where a_1, a_2, a_3, ... are digits. If with the digits in the even places we form the number x, and with the digits in the uneven places we form the number y, we have a reciprocal correspondence between one decimal fraction and two other decimal fractions. But two decimal fractions of different form, such as 0.0999 ... and 0.1000 ..., may have the same value; and the correspondence between the number t and the numbers x and y is not continuous. If we decompose the square of side 1 into 100 squares of side $1/10$, then if t goes from the values 0.0900 ...$^\sqcap$0.0999 ... to the values 0.1000 ...$^\sqcap$0.1999 ..., the point (x, y) goes from the last square in the first column to the first square in the second column, and these two squares are not adjacent.

We place the partial squares so that they will be adjacent. In base 2 enumeration, we suppose four partial squares to be in the order given in figure (a), and in base 3 in the order of figure (b).

Now we divide every partial square into other squares, and so on *ad infinitum.* Figure (c) represents the succession of 16 squares in base 2; figure (d) the succession of 81 squares in base 3.

If we represent by the sign ∩ the succession $\begin{smallmatrix} 1 & 2 \\ 0 & 3 \end{smallmatrix}$, or figure (a), then figure (e) represents the succession of 64 squares in base 2.

$$
\text{(a)} \quad \begin{matrix} 1 & 2 \\ 0 & 3 \end{matrix} \qquad \text{(b)} \quad \begin{matrix} 2 & 3 & 8 \\ 1 & 4 & 7 \\ 0 & 5 & 6 \end{matrix}
$$

(a)   (b)   (c)   (d)   (e)

In my article 'Sur une courbe qui remplit toute une aire plane,' I gave the analytic expression for the continuous correspondence between the real number t and the complex number $(x; y)$.

See also Hilbert, *Math. Ann.*, 38 (1891), 459; Cesaro, *Bull. sci. math.*, 21 (1897), 257; Moore, *Trans. Amer. Math. Soc.* (1900), 72; Lebesgue, *Leçons sur l'intégration* (Paris, 1904), p. 45.

XI

On some singular curves (1890)*

In all his work in analysis Peano continued to call for simplicity and rigour. His publication of counterexamples to commonly held notions had a large influence in the movement for greater rigour, especially in the teaching of calculus. The space-filling curve was a spectacular example. The following selection is typical of many such notes.

K.G.C. von Staudt, in §11 of his *Geometrie der Lage* (1847),[1] treats synthetically several properties of the tangent and osculating plane to a general curve. These properties are proved in infinitesimal calculus with Taylor's formula; hence we suppose that the functions can be developed by this formula. Staudt, however, does not subject the curves which he considers to any condition except continuity. I propose to show by examples that these restrictive conditions are necessary, and hence that the propositions stated by Staudt are not rigorously true; and I publish these examples, which I believe to be new, with the hope of persuading someone to make the propositions and proofs of Staudt rigorous, while still treating them synthetically.

Staudt, in no. 144, states a proposition which, in ordinary language, would read: 'If P and P' are two points of a planar curve, having tangents at every point, as P' tends toward P, the point of intersection of the tangents t and t' at the points P and P' has the point P as limit.'

We can, however, give curves for which, as P' tends toward P, the point tt' does not tend to any limit. Consider the equation $y = x^2 \sin(1/x)$, and let $y = 0$ for $x = 0$. Then y is a continuous function of x. Therefore the equation represents a continuous curve, passing through the origin and, since $\lim_{x=0}(y/x) = 0$, having as tangent there the x-axis. Considering another point with abscissa x, we have $dy/dx = y' = 2x \sin(1/x) - \cos(1/x)$,

---

* 'Sopra alcune curve singolari,' *Atti Accad. sci. Torino*, 26 (1890–1), 299–302 [29].
1 [Peano's remarks were prompted by the publication in 1889 of Mario Pieri's Italian translation of von Staudt's work.]

and the point of intersection of the tangent there with the tangent at the origin will have as abscissa

$$x - \frac{y}{y'} = x - \frac{x^2 \sin(1/x)}{2x \sin(1/x) - \cos(1/x)},$$

and 0 as ordinate. As we make x tend toward 0, the abscissa $x - y/y'$ does not tend toward any limit, but has $-\infty$ and $+\infty$ as oscillating extremes. Hence the point of intersection of the tangent at P with the tangent at O does not tend to any limit; but if we arbitrarily choose a point A on the x-axis, in every arc of the curve, no matter how small, having O at one end, we shall always have points at which the tangent passes through A. For this curve, y', as x tends to zero, does not tend to a limit, but has the oscillating extremes -1 and $+1$; hence the tangent at P does not tend to any limit as P tends to O.

In the example just given the curve is intersected by the x-axis at an infinite number of points in a neighbourhood of the origin. The curve

$$y = x^2 \left(2 + \sin\frac{1}{x}\right)$$

is intersected by every straight line at a finite number of points; and has the same properties as the preceding example.

We can give curves for which, as P' tends toward P, the tangent t' has t as limit without, however, the point tt' tending toward any limit. Consider the curve

$$y = x^3 \sin\frac{1}{x}.$$

It passes through the origin, and has the x-axis as tangent there. We have

$$y' = 3x^2 \sin\frac{1}{x} - x \cos\frac{1}{x}, \qquad \text{and } \lim_{x=0} y' = 0;$$

so that the tangent at the point with abscissa x has as limit the tangent at the origin. But the abscissa of the point of intersection of the two tangents

$$x - \frac{y}{y'} = x - \frac{x^2 \sin(1/x)}{3x \sin(1/x) - \cos(1/x)}$$

does not tend to any limit as x tends to zero, but has the oscillating extremes $-\infty$ and $+\infty$.

We can give curves for which, as P' tends toward P, the tangent t' has t as limit, and the point tt' tends toward a limit, without the limit being P. Consider the curve whose equation is

$$y = x^{2/3} + x^2 \left(\sin^{2/3} \frac{1}{x^3} - \cos^{2/3} \frac{1}{x^3} \right).$$

It passes through the origin, and has the y-axis as tangent there. We have:

$$y' = \frac{2}{3} x^{-1/3} - \frac{2}{x^2} \cdot \frac{\cos^{4/3} \frac{1}{x^3} + \sin^{4/3} \frac{1}{x^3}}{\cos^{1/3} \frac{1}{x^3} \sin^{1/3} \frac{1}{x^3}}$$

$$+ 2x \left(\sin^{2/3} \frac{1}{x^3} - \cos^{2/3} \frac{1}{x^3} \right).$$

The third term in y' tends to 0, and the sum of the first two terms may be written:

$$\frac{1}{x^2} \left(\frac{2}{3} x^{5/3} - 2 \frac{\cos^{4/3}(1/x^3) + \sin^{4/3}(1/x^3)}{\cos^{1/3}(1/x^3) \sin^{1/3}(1/x^3)} \right).$$

In the expression between parentheses, the first term tends to zero. In the second term the numerator is between 1 and $^3\sqrt{2}$, and the denominator is, in absolute value, less than 1; hence the second term is, in absolute value, greater than 2. Multiplying by $1/x^2$, which tends to ∞, we deduce that $\lim_{x=0} y' = \infty$. We also have $\lim_{x=0} xy' = \infty$. Hence the tangent at the point with abscissa x has the y-axis as limit, i.e., the tangent at the origin. The point of intersection of the two tangents, which has 0 as abscissa and $y - xy'$ as ordinate, tends toward the point at infinity on the y-axis as x tends to zero, since $\lim y = 0$ and $\lim xy' = \infty$.

Analogously, admitting only the continuity of a curve in space, if P and P' are two points on it, t and t' the tangents, π and π' the respective osculating planes, we may not give, as Staudt does, the propositions:

'The plane π is the limit of the plane passing through t and parallel to t'.' (no. 146)

'The plane π is the limit of the plane Pt'.' (no. 148)

'The straight line t is the limit of the straight line $\pi\pi'$.' (*ibid.*)

'The point P is the limit of the point $t\pi'$.' (*ibid.*)

'The point P is the limit of the point $t'\pi$.' (*ibid.*)

XII
The principles of mathematical logic (1891)*

In 1891 Peano founded the Rivista di matematica and contributed to the first volume five original articles, observations on another article, a reply to a declaration, an open letter, and four book reviews. The following selection occupies the first ten pages of this journal and summarizes Peano's work, to that date, in mathematical logic. The reader can test for himself the influence of Peano on the development of logic and logical notation by noting the ease with which he reads this paper. It is remarkable that it appeared only three years after Peano's first publication in logic.

Leibniz[1] announced several analogies between the operations of algebra and those of logic, but it is only in this century, through the work of Boole, Schröder, and many others,[2] that these relations have been studied to such an extent that deductive logic has become, like ordinary algebra, the theory of quaternions,[3] etc., a part of the calculus of operations.

---

* 'Principii di logica matematica,' *Rivista di matematica*, 1 (1891), 1–10 [31].

1 Leibniz was concerned several times with this question. Here are some sentences from his *Dissertatio de arte combinatoria* (Leipzig, 1666), no. 90: 'Truly, once the tables or categories of our involved art have been established, great things will come forth. For primitive terms will be designated by notes or letters as in an alphabet since from a mastery of them all other things are to be established ... If these are properly and clearly established, then this language will be universal, equally easy and common, and able to be read without any sort of dictionary, and at the same time a basic knowledge of all things will be assimilated.' [Peano quoted this passage in the original Latin. The translation given here was furnished by the Reverend Dennis C. Kane, O.P.]

2 The principal work of Boole is entitled *An Investigation of the Laws of Thought* (424 pp., London, 1854). This book, which is rare in Italy, may be found in the V. E. Library of Rome.

  The latest work of Schröder is *Algebra der Logik* (Leipzig, 1890), of which the first volume of 720 pages has appeared. I refer to this work because of its numerous citations, limiting myself here, for Italian readers, to mentioning my *Calcolo geometrico, preceduto dalle operazioni della logica deduttiva* (Turin, 1888) and the excellent article: Nagy, 'Fondamenti del calcolo logico,' *Giornale di matematiche*, 28 (1890), 1–35.

3 The analogy between the calculus of logic and that of quaternions lies in the fact that

One of the most notable results reached is that, with a very limited number (7) of signs, it is possible to express all imaginable logical relations, so that with the addition of signs to represent the entities of algebra, or geometry, it is possible to express all the propositions of these sciences.[4]

In the present note I shall explain these theories in a summary way, with the purpose of interesting the reader in this type of study, in itself most interesting, and of preparing for myself a tool which is almost indispensable for future research.

1 DEDUCTION AND CONJUNCTION

In this section the letters a, b, ... indicate any propositions whatever.

The expression $a \supset b$ means 'from a may be deduced b,' and it may be read 'if a is true, then b is true,' or 'if a, then b,' as well as in other ways.[5]

The expression $a = b$ means that the propositions a and b are equivalent, or that from the first the second may be deduced, and vice versa.

The simultaneous affirmation of several propositions a, b, c, ... is indicated by writing them one after another abc... This simultaneous affirmation is called *conjunction* or *logical multiplication*. We have:

1. $ab = ba$.
2. $(ab)c = a(bc) = abc$.

the symbols of each science satisfy laws which are special and analogous to, although not identical with, those of ordinary algebra.

I believe it opportune to give the words of Tait (*An Elementary Treatise on Quaternions* [1867], p. 50, footnote): 'It is curious to compare the properties of these quaternion symbols with those of the Elective Symbols of Logic, as given in Boole's wonderful treatise on the *Laws of Thought*; and to think that the same grand science of mathematical analysis, by remarkably similar processes, reveals to us truths in the science of *position* far beyond the powers of the geometer, and truths of deductive reasoning to which unaided thought could never have led the logician.'

4 I reached this result in my booklet: *Arithmetices principia, nova methodo exposita* (Turin, 1889). I continued applying these methods in the notes: *I principii di geometria, logicamente esposti* (Turin, 1889); 'Les propositions du cinquième livre d'Euclide, réduites en formules,' *Mathesis*, 10 (1890), 73–5; 'Démonstration de l'intégrabilité des equations différentielles ordinaires,' *Math. Ann.*, 37 (1890), 182–228.

It thus results that the question proposed by Leibnitz has been completely, if not yet perfectly, resolved; or, to use the words of Schröder (*Algebra*, p. 710): 'It is most striking to discover ... an immense set of geometrical theorems, together with their proofs, ... presented solely in the language of signs. ... It appears from this that the ... ideal of a pasigraphy for the purposes of science has already been realized to a quite considerable extent.' [In this passage (translated here from the original German) Schröder was commenting on Peano's *I principii di geometria, logicamente esposti*.]

5 To indicate that the proposition b is a consequence of a we may write 'b C a,' where the sign C is the initial of the word *consequence*. Then one can, at convenience, indicate the same relation by interchanging the two members and inverting the sign C, as is done with the signs > and < ; so that this proposition, written '$a \supset b$,' always

These identities express the commutative and associative properties of logical multiplication, analogous to those of algebraic multiplication.

3. $aa = a.$

This identity has no analogue in algebra.[6]

In order to separate the various propositions one from another, we may use parentheses as in algebra. The same result is gained with greater simplicity (and without the confusion of parentheses in algebraic formulas) by using a convenient system of dots. The signs of the system of dots are . : ∴ :: etc. To read a formula divided by dots, take together first all the signs not separated by a dot, then those separated by one, then those by two, and so on.[7]

We have:

4. $a = b . = . b = a.$

'The proposition $a = b$ is equivalent to $b = a.$'

5. $a = b . = : a \supset b . b \supset a.$

'Two propositions a and b are equivalent when the second may be deduced from the first, and vice versa.'

The following formulas represent various kinds of *syllogisms*:

6. $a \supset b . b \supset c : \supset . a \supset c.$

7. $a = b . b \supset c : \supset . a \supset c.$

8. $a \supset b . b = c : \supset . a \supset c.$

9. $a = b . b = c : \supset . a = c.$

The sorites has the form:

10. $a \supset b . b \supset c . c \supset d : \supset . a \supset d.$

means 'b is a consequence of a,' or 'a has b for a consequence,' or 'from a we may deduce b.'

The sign of deduction has been given diverse forms. Many English authors write ∴ . C.S. Peirce ('On the Algebra of Logic,' *American Journal of Mathematics*, 3 [1880], 15) writes —≺. Schröder [*Algebra*] adopts a sign derived from the two, = and <. MacColl ('The Calculus of Equivalent Statements,' *Proceedings of the London Mathematical Society*, 10 [1878], p. 16) writes $a : b$. Frege (*Begriffsschrift*, Halle a. S., 1879), instead of $a \supset b$, writes

$$\vdash\!\!\!\!-\!\!\begin{array}{c} b \\ a \end{array}$$

Some authors do not introduce any sign for deduction, since it can also be expressed by a combination of other signs.

6 The law represented by this formula is due to Jevons (*The Principles of Science* [London, 1883]), and is called *the law of simplicity*.

7 Thus we already find $d.uv$ and $du.v$ in place of $d(uv)$ and $(du)v$. This system of dots has a certain analogy with the notation proposed by Leibniz (*Math. Schriften*, VII [1863], p. 55).

We have:

11. $a \supset b . \supset . ac \supset bc.$

12. $a = b . \supset . ac = bc.$

13. $a \supset b . c \supset d : \supset . ac \supset bd.$

14. $a = b . c = d : \supset . ac = bd.$

These formulas say that to the two members of a deduction or a logical equivalence may be joined the same proposition, and that two deductions or two equivalences may be joined, member to member.

2 SINGULAR PROPOSITIONS; CLASSES

The names which we shall adopt sometimes represent individuals (proper names) such as $1, 2, \frac{1}{4}, \sqrt{2}, \ldots$; and sometimes classes (common names or adjectives) such as *number, polygon, equilateral,* etc.

The expression $a = b$, where a and b are individuals, indicates their identity, or that a and b are two names given to one and the same individual. If a and b are classes, this expression indicates that the two classes coincide, or that every a is a b, and vice versa. The meaning of this expression has already been explained if a and b are propositions.

To indicate the singular proposition 'x is an individual of the class s,' we shall write[8]

$$x \, \varepsilon \, s$$

and the sign ε may be read *is*, or *is a*, or *was*, or *will be*, according to grammatical rules, but its meaning is always that explained.

For brevity we shall write $x, y, z \, \varepsilon \, s$ to indicate that x, y, z belong to s, or

$$x, y, z \, \varepsilon \, s . = : x \, \varepsilon \, s . y \, \varepsilon \, s . z \, \varepsilon \, s.$$

To indicate the universal proposition 'every a is a b,' or 'the class a is contained in b,' we shall write $a \supset b$. Hence the sign $\supset$ is read differently (*from ... may be deduced*, or *is contained in*) according to whether it stands between two propositions or between two classes – but its properties are the same.

If a and b are two classes, we shall indicate by ab the set of entities which are at the same time in a and in b, and analogously for abc, etc. Where, however, there is danger of confusion, we shall write $a \cap b$ and $a \cap b \cap c$ in place of ab and abc.

All the formulas of the preceding section hold true when $a, b, \ldots$ represent classes.

---

8 The sign ε is the initial of $\dot{\varepsilon}\sigma\tau\iota$.

3 APPLICATIONS

The signs ε, =, ⊃ allow us to express a large number of logical relations. Hence, with the introduction of symbols to indicate the individuals, the classes, the operations, and the relations of a science, we are already able to state several propositions completely. Let us take for example algebra, where we already have the symbols 1, 2, ... to represent individuals; $+, -, \times$, etc. for operations; $>, <, =, ...$ for relations; and let us introduce some signs to represent classes that often occur. We shall write:

N in place of 'positive integral number'
n   ”   'integral number'
R   ”   'positive rational number'
r   ”   'rational number'
Q   ”   'positive real number' or 'positive quantity'
q   ”   'real number or quantity,' which we shall simply call *number.*

The advantage of these symbols lies not only in their brevity, but in their exact meaning, and in the possibility of putting them into formulas. We have:

1. $a, b \, ε \, q \, . \, ⊃ \, . \, a + b \, ε \, q.$

'If a and b are two numbers, then $a + b$ is also a determined number.'

2. $a, b \, ε \, q \, . \, b \gtrless 0 : ⊃ \, . \, a/b \, ε \, q.$

'If a and b are two numbers, of which the second is not zero, then a/b represents a determined, finite number.'

3. $a, b \, ε \, q \, . \, ⊃ \, . \, a \times b = b \times a.$

'If two numbers are indicated by a and b, then we have etc.'

4. $m, n \, ε \, N \, . \, a \, ε \, q : ⊃ \, . \, a^{m+n} = a^m a^n.$

'If m and n are positive integers and a is a real number, then we have etc.'

5. $m, n \, ε \, q \, . \, a \, ε \, Q : ⊃ \, . \, a^{m+n} = a^m a^n.$

'If m and n are two real numbers, and a is a positive number, then we have etc.'

6. $a, b, c \, ε \, q \, . \, a < b : ⊃ \, . \, a + c < b + c.$
7. $m, n \, ε \, Q \, . \, m < n : ⊃ \, . \, (1 + 1/m)^m < (1 + 1/n)^n.$

The deductions from one relation to another may be stated analogously.

8. $a, b, x, y \, \varepsilon \, q \, . \, \supset \, \therefore \, x + y = a \, . \, x - y = b : \, = \, : 2x = a + b \, .$
$2y = a - b.$

'If a, b, x, y are numbers, the system of equations

$$x + y = a, \qquad x - y = b$$

is equivalent to the system

$$2x = a + b, \qquad 2y = a - b.'$$

9. $x, y \, \varepsilon \, q \, . \, \supset \, \therefore \, x^2 + y^2 = 0 \, . \, = \, : x = 0 \, . \, y = 0.$

'If x, y are (real) numbers, then $x^2 + y^2$ is zero if and only if both x and y are zero.'

The relations between equations and propositions may be expressed in an analogous way.

4 THE SIGNS $-$, $\cup$, Λ [9]

Let a be a proposition. By $-a$ we understand the negation of a. If a, b, c represent propositions, we have:

1. $-(-a) = a.$ 'Two negations make an affirmation.'
2. $a = b \, . \, = \, . \, -a = -b.$
3. $a \supset b \, . \, = \, . \, -b \supset -a.$

'The proposition "from a follows b" is equivalent to "from not b follows not a."'

For ease in writing, sometimes instead of writing the sign $-$ before the whole proposition, we shall write it before the sign of the relations ε, $=$, etc.:

4. $-(x \, \varepsilon \, s) \, . \, = \, . \, x - \varepsilon \, s.$
5. $-(x = y) \, . \, = \, . \, x - = y.$

9 The sign of negation $-$, in the form used here, is substantially due to Boole.

Instead of $-a$, Schröder [*Algebra*] writes a_1; Jevons [*Principles*] writes A; MacColl ('Calculus') a'.

Instead of $a \cup b$ Leibniz has $a \, \hat{v} \, b$ (where v is the initial of *vel*), Jevons uses $a \, . | \, . \, b$, and the majority of authors write $a + b$.

Dedekind (*Was sind und was sollen die Zahlen?* [1888]) instead of $a \cap b$ and $a \cup b$ writes $G(a, b)$ and $M(a, b)$, notations only slightly different from that already used by Cantor in his studies on *Mannigfaltigkeiten* (*Math. Ann.*, vol. 15 and following).

One could use the sign V, initial of *vero* [true], to indicate an identically true proposition, and, when dealing with classes, to indicate the *universal* class. This sign was used by Peirce ['Algebra']; it is the identity for logical multiplication. To indicate the absurd, or the null class, we shall use the sign Λ, i.e., the preceding letter turned upside down. We shall not introduce the sign V, which corresponds by duality to Λ, because we do not need it.

Instead of the signs Λ and V, the majority of authors write 0 and 1, or derived signs. It is, however, useful to distinguish the identities of the algebraic operations from those of the logical operations.

If a, b are propositions, we shall indicate by $a \cup b$ the affirmation of the truth of at least one of a and b – that is, either a is true or b is true. The operation $\cup$ is called *logical addition*.

We have:

6. $-(ab) = (-a) \cup (-b)$.

'To deny that both a and b are true at the same time is equivalent to affirming that a is not true or b is not true,' or 'the negation of a product is the sum of the negation of its factors.'

7. $-(a \cup b) = (-a)(-b)$.

'To deny that at least one of a and b is true is equivalent to affirming that both a and b are false.'

We have:

8. $a \cup b = b \cup a$, $a \cup (b \cup c) = (a \cup b) \cup c = a \cup b \cup c$,
    $a \cup a = a$,

which formulas express the commutative and associative properties of logical addition, analogous to those seen in §1.

9. $a(b \cup c) = ab \cup ac$,

which expresses the distributive property of logical multiplication with respect to logical addition, analogous to the algebraic $a(b + c) = ab + ac$.

EXAMPLES We have

$$x, y \, \varepsilon \, q \, . \, \supset \, \therefore \, xy = 0 \, . \, = \, : x = 0 \, . \, \cup \, . \, y = 0.$$

'Let x and y be two numbers. To say that their product is zero is the same as saying that one of the factors is zero.' If we take the negative of each member of the equality, then according to rules 2 and 7, it may also be written:

$$x, y \, \varepsilon \, q \, . \, \supset \, \therefore \, xy - = 0 \, : \, = \, . \, x - = 0 \, . \, y - = 0.$$

'If the product of two numbers is not zero, then they are both different from zero, and vice versa.'

Finally, we shall use the sign Λ to indicate absurdity. Hence $ab = \Lambda$ says that the propositions a and b are contradictory. We have:

10. $a - a = \Lambda$, $a \Lambda = \Lambda$, $a \cup \Lambda = a$.

'To affirm and deny the same proposition is an absurdity.' 'If in a system of equations some are contradictory, the system is absurd.' ... We may note the analogy between Λ and 0, which are the identities for logical addition and algebraic addition.

11. $a \supset b . = . a - b = \Lambda.$

'Instead of saying that b may be deduced from a, we may say that the assertion of a and the negation of b is an absurdity.' Hence in every deduction $a \supset b$ we may transport the second member b into the first, preceding it with the sign $-$, and writing Λ second; and vice versa. Thus, for example, proposition 9 from §1,

$$a = b . b = c : \supset . a = c,$$

may be written, transporting $a = c$ to the first member,

$$a = b . b = c . a - = c : = \Lambda,$$

'the system of proposition a is equal to b, b is equal to c, and a is not equal to c, is absurd'; and transporting $b = c$ into the second member, it may be written in the form

$$a = b . a - = c : \supset . b - = c.$$

It results from this that we could do without the sign $\supset$, always reducing the second member to Λ. We shall keep it, however, for greater variety and for analogy with the common form of expressing the thought.

If a, b, ... represent classes, to the signs $-$, $\cup$, Λ we shall give the following meanings:

$-a =$ 'the non-a,'

$a \cup b =$ 'the set of individuals which are in a or b,'

$\Lambda =$ 'the null set.' Hence $a \cap b = \Lambda$ means 'no a is a b.'

Hence the sign Λ is read *absurd* or *null*, according to whether one is dealing with propositions or classes; in the two cases it has the same properties, as does the sign $\supset$.

Regarding the sign of deduction ($\supset$) between two propositions, let us note what further follows. When a and b are propositions containing variable letters x, y, ..., by $a \supset b$ we understand 'whatever $x, y, ...$ are, so far as they satisfy a, b will also be true.' Thus

$$x, y \, \varepsilon \, Q . x - = y : \supset . x + y > 2\sqrt{xy}$$

means: 'Whatever the two positive quantities x and y are, if they are not equal, then etc.' But sometimes we want to affirm the deduction relative to only one or more of the variable letters. To indicate this, we shall write as index to the sign $\supset$ the letters relative to which the deduction is made. Thus, if a and b are propositions containing the letter x, besides other letters, the expression $a \supset_x b$ means 'whatever x may be, from a is deduced b.' This proposition, then, ceases to be an absolute proposition, but is a condition among the remaining letters. Thus we have:

$$a, b, c \, \varepsilon \, q : x \, \varepsilon \, q \, . \supset_x . \, ax^2 + bx + c = 0 \, \therefore \, \supset :$$
$$a = 0 \, . \, b = 0 \, . \, c = 0.$$

'If a, b, c are three numbers and if whatever the number x, we have $ax^2 + \ldots = 0$, then the three given numbers are zero.'

Analogously, $a =_x b$ indicates that, whatever x may be, the propositions a and b are equivalent, or $a \supset_x b \, . \, b \supset_x a$.

$a =_x \Lambda$ means 'whatever the value of x, a is absurd' or 'there is no x which satisfies the condition a.'

CONCLUSIONS

The signs ε (*is*), $=$ (*is equal*), $\supset$ (*implies* or *is contained*), $\cap$ (*and*, usually indicated by juxtaposition), $\cup$ (*or*), $-$ (*not*), and Λ (*absurd or empty*) allow the expression of any logical relation.

It is convenient to give a name to these signs, and those written in the parentheses may be used. But these names, taken from common language, represent them only approximately – inasmuch as the signs always have the same meaning, but the words can have several. To translate an ordinary proposition into symbols, it is necessary to analyse it, to see the meaning of the various words, and to represent these meanings with equivalent symbols. It would be a great error to substitute the symbols ε, $\cap$, $\cup$, ... unthinkingly in place of the words *is, and, or,* ...

In the preceding pages, several properties of these symbols have been stated. But there are many, many more. We should note the *duality principle*, by which one logical identity may be obtained from another by exchanging the signs $\cap$ and $\cup$; and I should mention at least that Boole succeeded in resolving every system of equations containing one or more unknown classes, bound to known classes by the operations $\cup$, $\cap$, $-$, repeated an arbitrary number of times.

XIII
A generalization of Simpson's formula (1892)*

Peano concerned himself several times with both the theoretical and practical aspects of quadratures. In the following sample of his work we find a general formula which includes the well-known trapezoidal and Simpson's rules as particular cases. Remainder terms are also calculated for two other particular cases.

Notable among quadrature formulas are the trapezoidal rule:

(α) $\quad \int_a^b f(x)dx = \dfrac{b-a}{2}[f(a) + f(b)] + R,$

where

$$R = -\frac{(b-a)^3}{12} f''(u),$$

and Simpson's rule:

(β) $\quad \int_a^b f(x)dx = \dfrac{b-a}{6}\left[f(a) + 4f\left(\dfrac{a+b}{2}\right) + f(b)\right] + R,$

where

$$R = -\frac{(b-a)^5}{4!\,5!} f^{iv}(u),$$

such that u is an intermediate value between a and b.[1] The remainder in (α) is zero if $f(x)$ is an entire function of first degree, and in (β) the remainder is zero if $f(x)$ is an entire function of degree not higher than three.

---

* 'Generalizzazione della formula di Simpson,' *Atti Accad. sci. Torino*, 27 (1892), 608–12 [50].

1 I published this expression for the remainder in Simpson's formula in *Applicazioni geometriche del calcolo infinitesimale* [Turin: Bocca, 1887], p. 210. [This seems to have been the first time this expression for the remainder in Simpson's formula appeared. It was also discovered independently by A.A. Markov and published by him in 1889.]

Parallel to these formulas are those of Gauss. The analogue of (β) is

(β′) $\int_a^b f(x)dx = \frac{b-a}{2}\left[f\left(\frac{a+b}{2} - \frac{1}{\sqrt{3}}\frac{b-a}{2}\right)\right.$

$$\left. + f\left(\frac{a+b}{2} + \frac{1}{\sqrt{3}}\frac{b-a}{2}\right)\right] + R,$$

where

$$R = -\frac{(b-a)^5}{180\cdot 4!}f^{iv}(u),$$

and the remainder is zero for functions of degree not higher than three.

Comparing the formulas (β) and (β′), which may be considered as approximately equal, we see that it is simpler, in general, to calculate the three values

$$f(a), \quad f\left(\frac{a+b}{2}\right), \quad f(b)$$

that make up formula (β), than to calculate the two

$$f\left(\frac{a+b}{2} - \frac{1}{\sqrt{3}}\frac{b-a}{2}\right), \quad f\left(\frac{a+b}{2} + \frac{1}{\sqrt{3}}\frac{b-a}{2}\right)$$

that make up (β′). This explains why Simpson's formula (β) is used more than the corresponding formula (β′) of Gauss.

The formulas of Gauss give us an infinite succession, whereas the formulas for the trapezoidal rule and that of Simpson have, up to now, been isolated examples. I propose to present here an infinite succession of quadrature formulas, of which the first two are precisely (α) and (β).

For simplicity, let us suppose that the limits of integration are −1 and +1, since it is sufficient to make the substitution

$$x = \frac{a+b}{2} + \frac{b-a}{2}x'$$

to reduce the integral with limits a and b to this case.

The question that we propose is this: to determine the $n + 1$ coefficients $A_0, A_1, ..., A_n$ and the $n - 1$ values $x_1, x_2, ..., x_{n-1}$ between −1 and +1 such that the formula

(1) $\int_{-1}^{+1} f(x)dx = A_0 f(-1) + A_1 f(x_1) + ... + A_{n-1}f(x_{n-1}) + A_n f(1)$

holds, whenever $f(x)$ is an entire function of degree $2n - 1$.

The solution is the following. Let

(2) $\quad Y_n = (d/dx)^{n-1}(x^2 - 1)^n.$

Since the function $(x^2 - 1)^n$ has -1 and $+1$ as multiple roots of order n, its $(n - 1)$th derivative, Y_n, will have the simple roots $x_0 = -1$ and $x_n = 1$, and $n - 1$ roots $x_1, x_2, ..., x_{n-1}$ which are distinct and between -1 and $+1$. The coefficients $A_0, A_1, ...$ may be calculated with the formula

(3) $\quad A_r = \int_{-1}^{+1} \dfrac{(x - x_0) ... (x - x_{r-1})(x - x_{r+1}) ... (x - x_n)}{(x_r - x_0) ... (x_r - x_{r-1})(x_r - x_{r+1}) ... (x_r - x_n)} \, dx.$

Then formula (1) holds.

In fact, divide $f(x)$, an entire function of degree $2n - 1$, by Y_n, of degree $n + 1$. Let $\varphi(x)$ be the quotient and $\psi(x)$ the remainder. Then

(4) $\quad f(x) = \psi(x) + Y_n\varphi(x).$

The function $\psi(x)$ will be of degree n and $\varphi(x)$ of degree $n - 2$. If we give to x the $n + 1$ values $x_0, x_1, ..., x_n$, for which Y_n vanishes, we shall have

$$f(x_0) = \psi(x_0), \quad f(x_1) = \psi(x_1), \quad ..., \quad f(x_n) = \psi(x_n).$$

Hence the function $\psi(x)$, an entire function of degree n, whose value is known for $n + 1$ values of the variable, may be expressed by the interpolation formula of Lagrange:

(5) $\quad \psi(x) = \sum_r \dfrac{(x - x_0) ... (x - x_{r-1})(x - x_{r+1}) ... (x - x_n)}{(x_r - x_0) ... (x_r - x_{r-1})(x_r - x_{r+1}) ... (x_r - x_n)} f(x_r).$

From (4) we have:

(6) $\quad \displaystyle\int_{-1}^{+1} f(x)dx = \int_{-1}^{+1} \psi(x)dx + \int_{-1}^{+1} Y_n\varphi(x)dx.$

Now, from (5) and (3), we get

(7) $\quad \displaystyle\int_{-1}^{+1} \psi(x)dx = \sum_r A_r f(x_r).$

With regard to the second integral, by integration by parts we get

$$\int Y_n\varphi(x)dx = \int \varphi(x)\left(\frac{d}{dx}\right)^{n-1}(x^2 - 1)^n dx$$

$$= \varphi(x)\left(\frac{d}{dx}\right)^{n-2}(x^2 - 1)^n - \varphi'(x)\left(\frac{d}{dx}\right)^{n-3}(x^2 - 1)^n + ...$$

$$\pm \varphi^{(n-2)}(x)\cdot(x^2 - 1)^n \mp \int \varphi^{(n-1)}(x)\cdot(x^2 - 1)^n dx.$$

Setting -1 and $+1$ as limits, all the integrated terms in the second member vanish, because they contain the factor $x^2 - 1$; and since $\varphi(x)$ is of degree $n - 2$, $\varphi^{(n-1)}(x) = 0$, so that

(8) $\displaystyle\int_{-1}^{+1} Y_n\varphi(x)dx = 0.$

Substituting into (6) for the two integrals of the second member whose values are given by (7) and (8), we have formula (1), which is what we wished to prove.

Formula (1) is exact if $f(x)$ is an entire function of degree $2n - 1$ and approximate if $f(x)$ is an arbitrary function. To calculate the error R, such that

(9) $\displaystyle\int_{-1}^{+1} f(x)dx = \sum_r A_r f(x_r) + R,$

form the entire function $F(x)$ of degree $2n - 1$ which satisfies the $2n$ conditions:

$$F(x_0) = f(x_0), \quad F(x_1) = f(x_1), \quad F(x_2) = f(x_2), \quad \dots$$
$$F(x_{n-1}) = f(x_{n-1}), \quad F(x_n) = f(x_n),$$
$$F'(x_1) = f'(x_1), \qquad F'(x_2) = f'(x_2), \quad \dots, \quad F'(x_{n-1}) = f'(x_{n-1}).$$

We shall have, as we have seen,

(10) $f(x) = F(x) + (x - x_0)(x - x_1)^2(x - x_2)^2 \dots$
$$(x - x_{n-1})^2(x - x_n)\frac{f^{(2n)}(u)}{(2n)!}.$$

Integrating, we get

$$\int_{-1}^{+1} F(x)dx = \sum_r A_r f(x_r),$$

so that

(11) $R = \displaystyle\int_{-1}^{+1} (x - x_0)(x - x_1)^2 \dots (x - x_{n-1})^2(x - x_n)\frac{f^{(2n)}(u)}{(2n)!}\,dx.$

Now take the factor $f^{(2n)}(u)/(2n)!$ across the integral sign.[2] We may do this since the remaining factor has the same sign throughout the interval of integration. We then have:

2 [Peano apparently assumes here that $f^{(2n)}(x)$ is continuous. If only boundedness is assumed, then the factor $f^{(2n)}(u)$ in equation (12) must be replaced by a number K such that K is intermediate among the values of $f^{(2n)}(u)$ for $0 \leqslant x \leqslant 1$.]

(12) $\quad R = \dfrac{f^{(2n)}(u)}{(2n)!} \displaystyle\int_{-1}^{+1} (x - x_0)(x - x_1)^2 \ldots (x - x_{n-1})^2(x - x_n)\,dx.$

Setting $n = 1$, we have the trapezoidal formula (α).
For $n = 2$, we have Simpson's formula (β).
For $n = 3$, after the calculations have been made, we have:

$$\int_{-1}^{+1} f(x)\,dx = \tfrac{1}{6}f(-1) + \tfrac{5}{6}f(-\sqrt{\tfrac{1}{5}}) + \tfrac{5}{6}f(\sqrt{\tfrac{1}{5}}) + \tfrac{1}{6}f(1) + R,$$

where

$$R = -\frac{2^5}{3 \cdot 5^2 \cdot 7 \cdot 6!}\, f^{(6)}(u),$$

which is zero for functions of degree less than six.
For $n = 4$ we have:

$$\int_{-1}^{+1} f(x)\,dx = \frac{1}{10}f(-1) + \frac{49}{90}f(-\sqrt{\tfrac{3}{7}}) + \frac{32}{45}f(0)$$

$$+ \frac{49}{90}f(\sqrt{\tfrac{3}{7}}) + \frac{1}{10}f(1) + R,$$

where
$$R = -\frac{f^{(8)}(u)}{2^2 \cdot 3^4 \cdot 5^2 \cdot 7^3} = -\frac{f^{(8)}(u)}{2778300}.$$

NOTE The same question has already been treated by the late D. Turazza, in his paper 'Intorno all'uso dei compartimenti diseguali nella ricerca del valore numerico di un dato integrale' (*Mem. R. Istituto Veneto*, 5 [1855], 277–98). I believe, however, that the expressions for Y_n and the remainders are new.

XIV
On linear systems (1894)*

In the following note Peano brought to the attention of a wider reading audience his earlier work in extending Taylor's formula to linear systems. This was in response to a long article of Emmanuel Carvallo, published in the same journal, which showed that he was unaware of Peano's results.

Mr Carvallo published in this journal [*Monatsh. Math.*, 2 [1891], 177–216, 225–66, 311–30] a long article on this subject, which is of the greatest importance. There are two omissions, however, that should be made good.

On page 187, there is assumed for e^φ, when φ is a linear system, the definition

$$e^\varphi = 1 + \varphi + \frac{\varphi^2}{2!} + \dots$$

and he adds: 'It remains to be shown that this development is convergent.' I gave this proof in my note 'Integrazione per serie delle equazioni differenziali lineari' (*Atti Accad. Torino*, 1887, and in French in *Math. Ann.*, 1888).

On page 228, the author, in treating the application of Taylor's formula to vectors, says: 'But this development must be indefinitely continued since we do not know how to give a formula for the remainder.' In my *Calcolo geometrico* (Turin, 1888) I gave several forms of the remainder R in the development

$$f(t + h) = f(t) + hf'(t) + \dots + \frac{h^n}{n!} f^{(n)}(t) + R,$$

where $f(t)$ is any complex whatever (imaginary number, vector, quaternion, linear system, etc.) which is a function of the real variable t. We have:

$$(1) \quad \lim_{h=0} \frac{R}{h^n} = 0,$$

* 'Sur les systèmes linéaires,' *Monatshefte für Mathematik*, 5 (1894), 136 [62].

that is, R is an infinitesimal, with h, of order higher than n, and this on the sole supposition of the existence, for the particular value of the variable t, of the derivatives given in the formula, without supposing their existence or continuity in a neighbourhood of the value considered.

$$(2) \quad R = \frac{h^{n+1}}{(n+1)!} m,$$

where m is an intermediate value among those of $f^{(n+1)}(t + \theta h)$,[1] as the real variable θ varies between 0 and 1. This corresponds to Lagrange's expression, but we may not say here that m is one of the values of $f^{(n+1)}(t + \theta h)$, as we could if $f(t)$ were a real function.

$$3) \quad R = \frac{h^{n+1}}{n!} \int_0^1 (1 - \theta)^n f^{(n+1)}(t + \theta h) d\theta,$$

just as for real functions.

I have not given here the proof of these formulas, and of the many others that are obtained by transforming them, since this is a very simple application of theorems in analysis generalized to number complexes of any order.

---

1 [I.e., the modulus of m is between the g.l.b. and the l.u.b. of the moduli of $f^{(n+1)}(t + \theta h)$.]

XV
Essay on geometrical calculus (1896)*

The following selection was written by Peano for 'colleagues, not students.' It is, therefore, on a somewhat more sophisticated level than the earlier Calcolo geometrico secondo l'Ausdehnungslehre di H. Grassmann. *In the brief historical introduction we learn that Peano, like many others, had difficulty reading Grassmann's work. He 'uncovered the power of the new method,' he tells us, 'only in examining its applications.' He then reconstructed the theory, and it was this reconstruction which formed the basis of his treatise of 1888.*

The geometrical calculus differs from Cartesian geometry in that whereas the latter operates analytically with coordinates, the former operates directly on the geometrical entities.

The first step in geometrical calculus was taken by Leibniz, whose vast mind opened several new paths in mathematics. The infinitesimal calculus was developed first, through the work of his colleagues and disciples. The geometrical calculus, which concerns us here, was developed in this century, although it is not yet sufficiently widespread. Mathematical logic, whose principles were clearly set forth by Leibniz, is only now rapidly developing and resolving the various difficulties which lie in its path. On some points Leibniz confined himself to giving only hints. Of geometrical calculus he spoke with emphasis in a letter to Huygens (8 September 1679) in which he explains its great advantages.

After Leibniz, passing over the geometrical interpretation of imaginaries by Argand, we find Möbius rather successfully dealing with the same question in his *Der barycentrische Calcul* (1827), which he applied to several questions, and which he constantly used in his *Die Elemente der Mechanik des Himmels* (1843). Contemporaneously, and by an independent way, Bellavitis explained his *Metodo delle equipollenze* (1854), whose beginnings are found in his work of 1832, and of which he made numerous applications. In 1844 H. Grassmann published his *Ausdehnungslehre*, a work seldom read or appreciated by his contemporaries, but which was later admired by

* 'Saggio di calcolo geometrico,' *Atti Accad. sci. Torino*, 31 (1895–6), 952–75 [90].

numerous scientists, and with which we shall be especially concerned in the present article.

To end this brief historical sketch, Hamilton created by an entirely independent way the theory of quaternions, which is a new method of geometrical calculus. A hint of this theory was published in 1843, and the complete exposition was given in *Lectures on Quaternions* (1853). The work of Hamilton was well received by his contemporaries, this good fortune being due in part to his clarity of exposition, and to the felicitous nomenclature introduced. Quaternions have been used in many works and treatises, in both pure and applied mathematics, among them being, for example, the *Treatise on Electricity and Magnetism* (1873) of Maxwell. But, today, many are seeking to simplify the theory of quaternions (Macfarlane), while others are returning to the ideas of Grassmann, or seeking to combine the various methods of geometrical calculus.

Indeed these various methods of geometrical calculus do not at all contradict one another. They are various parts of the same science, or rather various ways of presenting the same subject by several authors, each studying it independently of the others.

It follows that geometrical calculus, like any other method, is not a system of conventions, but a system of truth. In the same way, the methods of indivisibles (Cavalieri), of infinitesimals (Leibniz), and of fluxions (Newton) are the same science, more or less perfected, explained under different forms.

The theory of Grassmann is judged these days, by the various authors who have re-explained and applied it, with the highest praise. The reader can consult the bibliographic list on the *Ausdehnungslehre* of H. Grassmann published in the *Rivista di matematica*, 5 (1895), 179–82, and especially the recent and very important little work of V. Schlegel, 'Die Grassmann' sche Ausdehnungslehre' [*Zeitschrift für Mathematik und Physik*, 41 [1896], Historisch-literarische Abtheilung, 1–21, 41–59.]

There must be a reason why this work has taken so long to become known and has met such difficulties. In my opinion, it lies in the form of exposition, a nebulous and metaphysical form, far from the usual language of mathematicians, and which from the beginning, instead of attracting the reader, tires him out and alienates him. Indeed, in studying this work, I uncovered the power of the new method only by examining its applications, especially those published in his 'Kurze Uebersicht über das Wesen der Ausdehnungslehre,' *Archiv der Mathematik und Physik*, 6 (1845), 337–50 (*H. Grassmann's Werke*, I, 297–312).

Starting out from these applications, I found it possible to reconstruct the theory and give definitions of the entities introduced, using only elementary geometry. I published this theory, giving numerous applications, in my *Calcolo geometrico secondo l'Ausdehnungslehre di H. Grassmann* (Turin, 1888). The same definitions were immediately adopted by Mr Carvallo in the article 'La méthode de Grassmann' (*Nouvelles annales de mathématiques*, [3], 11 [1892], 8–37), an article notable for its clarity and simplicity of exposition. Subsequently, in my *Lezioni di analisi infinitesimale* (1893), I expressed the propositions of this theory in the symbols of mathematical logic. Thus, through the work of various authors, the exposition of Grassmann's method has been made very simple and its application to the various fields of mathematics has continued to expand.

The complete exposition of the geometrical calculus, supposing known only elementary mathematics, would necessarily be rather voluminous. But many procedures of this calculus are similar to processes introduced (often later in the development) in analytic projective geometry, in analysis, and in mechanics, and several theorems are known in these sciences under somewhat different forms. Therefore, in directing myself not to students, but to colleagues, I believe I can satisfy a desire manifested by several by briefly explaining the definitions and fundamental properties of the entities on which the geometrical calculus works, contrasting them with analogous entities which are considered in various branches of mathematics.

The entities introduced will be defined here, viz. the geometrical forms of first, second, third, and fourth degree, of which vectors, bivectors, and trivectors are particular cases; the equivalence relation, the only relation considered here; the operations of addition and multiplication, and the two operations indicated by the signs ω and $|$. This complete system of operations allows the treatment of all the questions of geometry. We can consider only a few of these operations and entities in detail.

1 TETRAHEDRONS

If A, B, C, D are points, $ABCD$ indicates the tetrahedron whose vertices are the given points. $ABCD = 0$ signifies that the four points lie in a plane.

For a non-zero tetrahedron we shall consider, following Möbius, its sense. The tetrahedron $ABCD$ will be said to be right-handed if a person with his head at A, his feet at B, and facing CD had C on his left and D on his right. The contrary case will be called left-handed.

The concept of sense of a tetrahedron, although rather simple, being reduced to either left or right, is not found in the books of Euclid; it must be explained by imagining a person situated in the manner indicated. Once the physical concept of sense of a tetrahedron has been introduced, it becomes possible to define the sense of the other entities we shall introduce.

Two tetrahedrons are said to be *equal* if they have the same size and the same sense. Tetrahedrons may be added and multiplied by positive or negative real numbers – the result is always a tetrahedron. Thus, if $t_1, ..., t_n$ are tetrahedrons, and $x_1, ..., x_n$ are real numbers, then $x_1 t_1 + ... + x_n t_n$ is a tetrahedron and this polynomial has all the properties of algebraic polynomials; i.e. the order of the terms may be inverted, and the multiplication of a number by a tetrahedron has the distributive property with respect to both factors.

The tetrahedron $ABCD$ changes sign if two vertices are interchanged, i.e. (Möbius, *Werke*, p. 41):

$$ABCD = -BACD = -ACBD = -ABDC.$$

By the ratio t/u of two tetrahedrons t and u the second of which is not zero, we understand the real number which multiplied by u gives t. This ratio is also said to be the number which measures t when u is the unit of measure. In many problems, instead of tetrahedrons we may speak of the numbers which measure them.

The tetrahedron $ABCD$ will also be said to be the product of four points A, B, C, D, or of the point A by the triangle BCD, or of the line AB by the line CD, or of triangle ABC by the point D. This product does not have the commutative property, but the anticommutative property, since interchanging two vertices produces a change in sign. Later we shall see reasons which justify the name of product.

2 GEOMETRICAL FORMS

Let $x_1, x_2, ..., x_r$ be real numbers, and indicate points by capital letters. We define:

$$(x_1 A_1 + ... + x_r A_r)BCD = x_1 A_1 BCD + ... + x_r A_r BCD,$$
$$(x_1 A_1 B_1 + ... + x_r A_r B_r)CD = x_1 A_1 B_1 CD + ... + x_r A_r B_r CD,$$
$$(x_1 A_1 B_1 C_1 + ... + x_r A_r B_r C_r)D = x_1 A_1 B_1 C_1 D + ... + x_r A_r B_r C_r D.$$

The first members of these equalities have no significance up to this point; we attribute to them the value represented by the second members, which are sums of tetrahedrons.

A *form of the first degree* is an expression of the form

$$x_1 A_1 + ... + x_r A_r,$$

i.e., the set of points $A_1, ..., A_r$ with the coefficients or masses $x_1, ..., x_r$.

A *form of the second degree* is an expression of the form

$$x_1 A_1 B_1 + ... + x_r A_r B_r.$$

A *form of the third degree* is an expression of the form:

$$x_1 A_1 B_1 C_1 + \ldots + x_r A_r B_r C_r.$$

A *form of the fourth degree* is a sum of tetrahedrons, which is then reducible to a single tetrahedron.

Multiplying a form of the first degree by three points, or a form of the second degree by two points, or a form of the third degree by one point, yields tetrahedrons.

A form s of $\begin{Bmatrix} \text{first} \\ \text{second} \\ \text{third} \end{Bmatrix}$ degree is said to be zero, and we write $s = 0$, if multiplying by $\begin{Bmatrix} \text{three points} \\ \text{two points} \\ \text{one point} \end{Bmatrix}$, chosen arbitrarily, always yields a zero product.

Two forms s and s' of $\begin{Bmatrix} \text{first} \\ \text{second} \\ \text{third} \end{Bmatrix}$ degree are said to be equal, and we write $s = s'$, if multiplying by $\begin{Bmatrix} \text{three} \\ \text{two} \\ \text{one} \end{Bmatrix}$ arbitrary points always yields equal products.

The definitions of the vanishing of a form, and of the equality of two forms, are fundamental in our theory. We must always return to them whenever any doubt arises as to the interpretation of a formula.

The product of a geometrical form of first degree by a triangle BCD is proportional, as the form varies, to the moment of this form with respect to the plane of the triangle, i.e., to the sum of the distances of the points of the form from the plane, multiplied by their respective masses. Hence, two forms of first degree are said to be equal when they have the same moment with respect to every plane.

If a line AB represents a force, the product $ABCD$ is proportional to the moment of that force with respect to the axis CD. Hence, two forms of second degree are said to be equal when they have the same moment with respect to every axis.

We see here the analogy of forms of the first degree with the theory of barycentres, and of forms of the second degree with the reduction of forces applied to a rigid body.

3 OPERATIONS ON THE FORMS

We have already defined *line* as the product AB of two points, and *triangle* as the product ABC of three points. But these words have a special significance here and are particular cases of forms of the second and third degrees. Hence the equality $ABC = A'B'C'$ means, according to the definitions given, that however the point D is taken we always have $ABCD = A'B'C'D'$, which is the same as saying that the two triangles lie in the same plane and have the same size and the same sense. Analogously, $AB = A'B'$ means that the two segments lie on the same line, have the same length and the same sense, and this not by definition, but as an immediate consequence of the definition.

We shall now give the following intuitive definitions of the sum and product of geometrical forms:

The sum of two forms of the same degree is the form obtained by writing the terms of the second after the terms of the first.

The product of a form of degree i by one of degree j, supposing that $i + j \leqslant 4$, is the sum of the products of every term of the first by every term of the second.

An immediate consequence of these definitions is that the geometrical calculus differs from the algebraic calculus in that:

1. We may multiply only two, or three, or four points: there is no form of degree higher than four.

2. We have $AB = -BA$, and so $AA = 0$.

Otherwise, the geometrical calculus has all the properties of the algebraic calculus of polynomials. Addition is commutative and associative, and multiplication is associative and distributive with respect to both factors. In every case we may substitute one form for its equal.

The coincidence of the two calculi constitutes the immense advantage of the method of Grassmann. It makes it possible to operate and reason with great savings of effort and memory, since in this new calculus we operate as in an already known calculus. This method responds then to the principle of least effort – a principle which holds not only in mechanics, but in pedagogy as well.

Conversely, if we attribute to $ABCD$ the significance just given it and if we wish to keep the rules of algebra just mentioned, then of necessity we obtain the calculus of Grassmann. This could serve as a definition of this calculus, but such a definition would be redundant. It is the equivalent of a set of propositions, some of which are definitions and the others their consequences.

4 VECTORS

Of the forms of first degree special mention must be made of the difference $B - A$ of two points, i.e., the set of two points A and B with the coefficients -1 and $+1$. Such a difference is said to be a *vector*.

Two vectors $B - A$ and $B' - A'$ are equal when, by the definitions given, however the points PQR are taken, we have

$$BPQR - APQR = B'PQR - A'PQR.$$

This condition is easily transformed into another: Two vectors are equal when they have the same length, are parallel, and have the same direction.

A and B are said to be the *origin* and *terminus* of the vector $B - A$.

The origin of a vector may be selected arbitrarily.

To sum a point A with a vector I, we determine the point B such that $B - A = I$. Transposing, we have $B = A + I$.

To sum two vectors I and J, we arbitrarily select a point A, and construct the point $A + I$ and then the point $A + I + J$; the vector $(A + I + J) - A$ is equivalent to $I + J$. It follows from this that the sum of two vectors is a vector.

Multiplying a vector by a real number results in a vector parallel to the first.

Given a form of first degree

$$x_1 A_1 + \ldots + x_n A_n$$

and a point O, we have the identity:

$$x_1 A_1 + \ldots + x_n A_n = (x_1 + \ldots + x_n)O + x_1(A_1 - O) + \ldots$$
$$+ x_n(A_n - O),$$

i.e., 'every form of first degree is reducible to an arbitrary point O with coefficient the sum of the coefficients of the form given, plus a vector.'

'Every form of first degree for which the sum of the coefficients is zero is reducible to a vector.'

'Every form of first degree for which the sum of the coefficients is not zero, divided by that sum, gives a point.'

This point is the *barycentre* of the given points with their respective masses. In this way the barycentric calculus appears. To be precise, the theory of forms of first degree coincides in both substance and notation with the calculus of Möbius. Möbius, however, merely gave hints in the case in which the form reduces to a vector, not bringing out the importance of this case.

The term *vector* was introduced by Hamilton. It corresponds exactly to Bellavitis's *segment*, but the first name, which excludes equivocation, is the one that is coming more and more into general use.

These authors considered the vector directly, without making it depend on forms of the first degree, assuming as definitions of equivalence and sum those properties which we have just explained. Indeed, the concept of vectors and their sum, or composition, is quite a bit older, since it already appears in velocities and forces, but the credit must always be given to these authors for having made us see how, on the basis of their composition, a geometrical calculus could be founded.

Both Bellavitis and Hamilton indicated by AB the vector with origin A and terminus B, which we indicate, following Grassmann, by $B - A$. Hamilton noted the advantage of indicating it by $B - A$, but he never made use of this notation.

According to our definitions, $B - A$ and AB are entirely distinct entities. Independently of this, the formulas of Hamilton (A, B, C are points)

$$BA = -AB, \qquad AB + BC + CA = 0$$

express what is given by Grassmann's

$$A - B = -(B - A), \qquad (B - A) + (C - B) + (A - C) = 0.$$

We see that the latter have the form of algebraic identities, whereas the former have a different form, requiring a new effort to remember them. Thus considered, the calculus of Grassmann represents a greater economy over that of Hamilton.

Bellavitis introduced a sign to indicate the equipollence of two segments, so that for segments one has to consider equality and equipollence. Two equipollent segments may be substituted one for the other in certain formulas, which must be remembered. We, on the other hand, having only one sign of equality, may always substitute in any formula one entity for its equal. No rule could be simpler than this.

5 FORMS OF SECOND AND THIRD DEGREE

The theory of forms of second and third degree coincides with the theory of systems of forces applied to a rigid body, but in the former purely geometric concepts appear, and operations are performed on them with an algebraic algorithm.

Here are some samples. We have $A(B - A) = AB$, or every line is the product of a point and a vector, and vice versa. By vector of a line AB we mean the vector $B - A$. By vector of a form of second degree we mean the sum of the vectors of its terms.

We have the identity

$$AB_1 + AB_2 + \ldots + AB_n = A[A + (B_1 - A) + \ldots + (B_n - A)],$$

i.e., the sum of several lines having the same origin is a line having again the same origin and whose terminus is the point enclosed in brackets.

The product of two vectors is called a *bivector*. This is a form of second degree, which corresponds to a couple in mechanics. The sum of two bivectors is a bivector. Every form of second degree is reducible in an infinite number of ways to the sum of a line and a bivector.

The forms of third degree have no mechanical interpretation. We have the theorem: 'Every sum of triangles is reducible either to a single triangle or to the product of three vectors.'

The product of three vectors is called a *trivector*.

Adding a trivector to a triangle moves the triangle parallel to itself.

A brief exercise is sufficient to make one acquainted with this calculus, which differs from algebraic calculus only in that multiplication is anti-commutative.

The following observations may be helpful:

The vector $B - A$ is a form of first degree; the line AB is a form of second degree.

Two vectors $B - A$ and $B' - A'$ are equal when multiplication by three arbitrary points yields equal volumes.

From $AB = A'B'$ we deduce $B - A = B' - A'$, but not vice versa.

The bivector $(B - A)(C - A) = BC + CA + AB$ is a form of second degree; the triangle ABC is a form of third degree. Multiplying the bivector by the point A gives the triangle ABC.

Two bivectors $AB + BC + CA$ and $A'B' + B'C' + C'A'$ are equal when multiplication by two arbitrary points yields equal products; two triangles ABC and $A'B'C'$ are equal if multiplication by the same arbitrary point yields equal products. The equality $ABC = A'B'C'$ says that the two triangles lie in the same plane, have equal areas, and have the same sense. The equality $AB + BC + CA = A'B' + B'C' + C'A'$ says that the two traingles lie in parallel planes, have equal areas, and have the same sense.

From $ABC = A'B'C'$ we deduce $AB + BC + CA = A'B' + B'C' + C'A'$, but not vice versa.

A trivector is reducible to the form $(B - A)(C - A)(D - A)$ or, multiplying out,

$$BCD - ACD + ABD - ABC,$$

i.e., to the surface of the tetrahedron $ABCD$. Multiplying the trivector by an arbitrary point yields this tetrahedron. If two tetrahedrons are equal, then so are their trivectors, and vice versa.

Suppose we have in a fixed plane a form of second degree, i.e., a system of lines

$$A_1B_1 + A_2B_2 + \ldots + A_nB_n.$$

If the vector of this form is not zero, it is reducible to a single line CD, so that, however we choose P, in the same plane, we have

$$PA_1B_1 + PA_2B_2 + \ldots + PA_nB_n = PCD.$$

The construction of the line CD which results from our theory is identical with the construction of the resultant of the forces $A_1B_1 + A_2B_2 + \ldots + A_nB_n$ and of the transformation of a sum of triangles into one triangle, and, as a particular case, of the transformation of a polygon into a triangle, as is taught in graphical statics. But we have admitted the concept of area without analysing it. It is now easy to see that this transformation is based on the identity involving three vectors I, J, K:

$$(I + J)K = IK + JK,$$

which constitutes the theorem of Varignon.

The bivectors, or areas of the two members, may be decomposed into superposable parts. Hence the given polygon may be effectively decomposed into parts which, disposed differently, form the triangle said to be its equal. We thus have a way to resolve the question, much debated in recent years, whether polygons equal according to Euclid may be decomposed into superposable parts. This proof coincides substantially with that of Mr L. Gérard, 'Sur la mesure des polygones,' *Bull. de math. élém.*, 1 (1896), 100–2.

6 COORDINATES

Let A_1, A_2, A_3, A_4 be four forms of first degree whose product is not zero. Let them be reference forms. Now, every form of first degree may be reduced to the form

$$x_1A_1 + x_2A_2 + x_3A_3 + x_4A_4,$$

where $x_1, \ldots, x_4$ are numbers, which are said to be *coordinates* of this form.

Every form of second degree is a linear function of the six products obtained by taking the reference forms two by two, i.e., we may write

$$y_{12}A_1A_2 + y_{13}A_1A_3 + \ldots + y_{34}A_3A_4.$$

The six numbers $y_{12}, \ldots, y_{34}$ are said to be the coordinates of the form of second degree.

Every form of third degree is a linear function of the products obtained by taking the reference forms three by three, i.e., it may be equated to

$$z_1A_2A_3A_4 - z_2A_1A_3A_4 + z_3A_1A_2A_4 - z_4A_1A_2A_3.$$

The four numbers $z_1, \ldots, z_4$ are the coordinates of the form.

The coordinates of a form are easily expressed as ratios of tetrahedrons. Indeed, let

$$S = x_1 A_1 + x_2 A_2 + x_3 A_3 + x_4 A_4.$$

Multiply by $A_2 A_3 A_4$. In the second member only the first term will remain, whence, solving for x_1,

$$x_1 = \frac{S A_2 A_3 A_4}{A_1 A_2 A_3 A_4};$$

the other coordinates are obtained analogously.

Consider a form of second degree:

$$s = y_{12} A_1 A_2 + y_{13} A_1 A_3 + \dots$$

Multiply by $A_3 A_4$, and solve:

$$y_{12} = \frac{s A_3 A_4}{A_1 A_2 A_3 A_4}.$$

Forms of third degree may be dealt with analogously.

We may solve, with great simplicity, various problems dealing with coordinates, such as finding the coordinates of sums or products of forms of given coordinates. Indeed, it is sufficient to carry out the operations indicated.

For example, suppose we wish to calculate the volume of the tetrahedron which is the product of four forms of first degree with the coordinates $(x_1, \dots, x_4)$, $(y_1, \dots, y_4)$, $(z_1, \dots, z_4)$, $(t_1, \dots, t_4)$. Now, the given forms have the expressions:

$$x_1 A_1 + x_2 A_2 + x_3 A_3 + x_4 A_4,$$
$$y_1 A_1 + y_2 A_2 + y_3 A_3 + y_4 A_4,$$
$$z_1 A_1 + z_2 A_2 + z_3 A_3 + z_4 A_4,$$
$$t_1 A_1 + t_2 A_2 + t_3 A_3 + t_4 A_4.$$

Multiply out these polynomials. The product will consist of several terms. However, when two terms in the same vertical column are multiplied, their product is zero, so it is only necessary to multiply terms differing in row and column. One term of the product will be $x_1 y_2 z_3 t_4 A_1 A_2 A_3 A_4$; other terms are obtained by permuting the indices 1, 2, 3, 4. If we wish to find the common factor $A_1 A_2 A_3 A_4$ it is necessary to give to the coefficients the sign + or

— according as the number of the inversion is even or odd. We thus have the product

$$\begin{vmatrix} x_1 & x_2 & x_3 & x_4 \\ y_1 & y_2 & y_3 & y_4 \\ z_1 & z_2 & z_3 & z_4 \\ t_1 & t_2 & t_3 & t_4 \end{vmatrix} A_1 A_2 A_3 A_4,$$

and this by the definition of *determinant*.

We see the concept of determinant appearing here; its theory may be developed entirely by the general methods of Grassmann. We only mention this here, however, since it is our intention to speak of geometrical applications.

The reference elements A_1, A_2, A_3, A_4 may be any whatever. The case in which we take for reference elements the point O and three non-coplanar vectors I, J, K merits special attention. This system of coordinates is called *Cartesian*. Every form of first degree is reducible to the form

$$mO + xI + yJ + zK,$$

every point to the form

$$O + xI + yJ + zK,$$

and every vector to the form

$$xI + yJ + zK.$$

Every form of second degree is reducible to

$$lOI + mOJ + nOK + pJK + qKI + rIJ.$$

If the form represents a system of forces, the six coordinates of the form are called in mechanics the *characteristics* of the system.

The product of the form by itself yields

$$(lp + mq + nr)OIJK.$$

The vanishing of this quantity is the condition that the form be reducible to a line or to a bivector.

Every bivector is reducible to the form

$$pJK + qKI + rIJ.$$

7 APPLICATIONS TO ANALYTIC PROJECTIVE GEOMETRY

The geometrical forms considered up to this point are closely related to known entities. A non-zero form of first degree is reducible to a point with mass or to a vector. If it is multiplied by a non-zero number the position of the point, or the direction of the vector, is not altered. Hence a form of first degree determines a point or a direction, i.e., determines in every case a projective point. The four coordinates of the form of first degree are called the homogeneous coordinates of this point. If the coordinates are multiplied by the same number, the form is multiplied by this number, so that the projective point does not vary.

If the reference elements A_1, A_2, A_3, A_4 are any four forms, the coordinates are said to be *projective*; if they are four points, we have *barycentric* coordinates; if a point and three vectors, *Cartesian* coordinates.

A form of second degree, non-zero but such that its product with itself is zero, is reducible to a line or to a bivector. It determines, therefore, a straight line or a direction, i.e., it determines in every case a finite or infinite straight line.

A form of second degree, such that its product with itself is not zero, determines a *linear complex*.

A non-zero form of third degree is reducible either to a triangle or to a trivector. In the first case a finite plane is determined; the trivector corresponds to an infinite plane. The four coordinates of the form are the homogeneous coordinates of the plane.[1]

8 REGRESSIVE PRODUCTS

Consider two forms of first degree (or points) A and B, and a triangle (or plane) π. We wish to find the point of intersection of the line AB with the plane π. Since every point of the line has an expression of the form $xA + yB$, and this point of intersection must lie in π, we must have

$$(xA + yB)\pi = 0,$$

i.e.,

$$xA\pi + yB\pi = 0.$$

This equation is satisfied if we take x and y in proportion to the volumes $B\pi$ and $-A\pi$. Hence, if $[t]$ represents the number which measures the tetrahedron t relative to a fixed tetrahedron, the form

$$[B\pi]A - [A\pi]B$$

lies on the line AB and in the plane π. The point determined is the intersection of that line with plane.

---

1 A more complete development of projective coordinates, deducing them from geometrical calculus, is found in C. Burali-Forti, 'Il metodo del Grassmann nella geometria proiettiva,' *Rend. Circ. mat. Palermo*, 10 (1896), 177–95.

Now, we may show that this form, which is presented as a function of A and B, is really a function of the product AB alone, i.e., that it does not vary if we take in place of A and B other forms A' and B' such that $AB = A'B'$. For this reason we shall call the expression obtained the product of AB by π, and we shall write

$$AB \cdot \pi = [B\pi]A - [A\pi]B.$$

The product of a form of second degree by one of third degree is hence a form of first degree, fully determined. The point determined is the intersection of the line and the plane determined by the given forms.

The product of two forms of third degree (planes) is defined analogously to be a form of degree $3 + 3 - 4 = 2$ (a line), and the product of three forms of third degree to be a form of degree $3 + 3 + 3 - 8$.

These products, called regressive, retain the distributive property with respect to both factors, and indeed there are notable resemblances between these products and the progressive products first considered. Such a study is interesting for higher geometry, inasmuch as the method of Grassmann permits the indication by symbols of every construction obtained by projection and sectioning, and allows reasoning with the formulas. Thus, one may transform one construction into another and recognize the degree of the locus so defined. Many authors, who are mentioned in my *Calcolo geometrico*, have done this, and obtained notable results. But, these products being somewhat less simple than the other operations, let this mention suffice.

9 THE OPERATION ω ON FORMS

The case of the regressive product in which one factor is a fixed trivector ω, taken as the unit, is important, but since we do not wish to speak here of the regressive product, we make the following definitions:

If s is a form of first degree, by ωs we indicate its *mass*, i.e., the sum of its coefficients.

If s is a form of second degree, by ωs we indicate the *vector* of s, i.e., the sum of the vectors of its terms, so that, if A and B are points, $\omega(AB) = B - A$.

If s is a form of third degree, by ωs we indicate the *bivector* of s, so that if A, B, C are points, we have

$$\omega(ABC) = (B - A)(C - A) = BC + CA + AB.$$

Analogously, $\omega(ABCD)$ will indicate the trivector of the tetrahedron $ABCD$, i.e.,

$$BCD - ACD + ABD - ABC.$$

The operation ω is distributive, i.e., $\omega(s + s') = \omega s + \omega s'$, which says that in all calculations the sign ω acts as a constant factor. The operation ω carried out on a vector, or a bivector, or a trivector gives 0 as a result.

10 THE INDEX OPERATION ON VECTORS AND BIVECTORS

Let the *metre* be the unit measure of length.

By *modulus* of a vector *I* we understand its length measured in metres.

By *modulus* of a bivector *IJ* we understand the area of the parallelogram whose sides are the vectors *I* and *J*, measured in square metres.

The modulus of a vector, or a bivector, is therefore zero or a positive number.

Given a trivector *IJK*, in addition to its size (i.e., the volume of the parallelepiped constructed on the three given vectors, measured in cubic metres) we must also consider its sense. The trivector is said to be positive if the tetrahedron *OIJK* is right-handed.

For ease in writing we shall identify a trivector by the number which measures it, preceded by the appropriate sign. In other words, let ω be the trivector product of three vectors, each of length one metre, pairwise orthogonal, and such that the tetrahedron $O\omega$ is right-handed. Then in the ratio IJK/ω we shall suppress (when there is no danger of ambiguity) the denominator, and we shall simply write *IJK*.

The name *index* of a bivector *IJ*, and notation $|(IJ)$, is given to that vector *K* which is normal to *IJ*, in the sense which makes $IJK = KIJ$ positive, and whose modulus is equal to that of *IJ*. If $K = |(IJ)$, we also say that *IJ* is the index of *K*, and we write $IJ = |K$. Thus the index of a bivector is a vector, and vice versa.

If the bivector represents a pair of forces, the index is said in mechanics to be the *axis of moment* of the pair.

The operation $|$ is distributive, and in calculations this sign acts as a constant factor.

The operation $|$ determines that polarity which is called *absolute* by some authors. It allows the study of metric properties of figures. Here are a few results:

$$I|I = (\text{mod } I)^2,$$

i.e., the product of a vector by its index is equal to the square of its modulus. Grassmann abbreviated $I|I$ by I^2. We may, without ambiguity, write I^2, and read it 'the square of *I*,' with the understanding that we mean $I|I$ and not $II = 0$. In the preceding equation we may solve for mod *I*. The same things hold if, instead of *I*, we substitute a bivector.

The expression $I/(\text{mod } I)$ represents a vector of unit length, which is directed in the same sense as I.

Let I and J be two vectors of unit length. Then IJ is a bivector and $\text{mod}(IJ)$ is a number, which is said to be the *sine* of the angle between the two vectors. If I, J are not equal to the unit measure, we must first reduce them to have

$$\sin(I, J) = \frac{\text{mod}(IJ)}{(\text{mod } I)(\text{mod } J)}.$$

The sine of the angle between two vectors is a number in the interval from 0 to 1.

Let I be a vector and j a bivector whose moduli are unity. The number Ij is said to be the *sine* of their angle. Whatever the moduli of I and j, we have

$$\sin(I, j) = \frac{Ij}{(\text{mod } I)(\text{mod } j)}.$$

The sine of the angle between a vector and a bivector is a number in the interval from -1 to $+1$.

By cosine of the angle between two vectors I and J we mean the sine of I and $|J$, i.e.,

$$\cos(I, J) = \frac{I|J}{(\text{mod } I)(\text{mod } J)}.$$

The cosine of the angle between two vectors lies in the interval from -1 to $+1$. After being cleared of fractions, this formula says that the product of a vector by the index of another is equal to the product of their moduli by the cosine of the angle between them.

The product $I|J$ was named by Grassmann the *internal product* of the two vectors. He named the bivector IJ their *external product*. This internal product appears in mechanics, where it is the work done by the force I acting on a point which receives the displacement J, and in several tracts in mechanics a special sign has been proposed for it.

Thus the operations of elementary mathematics distance between two points, area of a triangle, sine, cosine of an angle, which operations are not distributive, but have complex properties, are expressed by the use of the external and internal products of Grassmann, on which one operates with rules more or less identical with algebraic rules.

11 APPLICATIONS TO CARTESIAN GEOMETRY

Let I, J, K be three vectors of unit length, pairwise orthogonal, and such that $IJK = +1$. We have

$$I|I = J|J = K|K = 1, \qquad I|J = I|K = J|K = 0,$$
$$|(JK) = I, \qquad |(KI) = J, \qquad |(IJ) = K. \tag{1}$$

A vector U with coordinates x, y, z has the expression

$$U = xI + yJ + zK. \tag{2}$$

We may multiply U by $|U$, i.e., make a square of it. This is equal to (mod $U)^2$. Developing the square in the second member by algebraic rules, taking account of identity (1), we have:

$$(\text{mod } U)^2 = x^2 + y^2 + z^2. \tag{3}$$

If the vectors I, J, K are not orthogonal, we would also have the terms $2xyI|J + ...$, where $I|J = \cos(I, J)$.

The condition that the point

$$P = O + xI + yJ + zK \tag{4}$$

lie in the plane of the triangle with coordinates a, b, c, d, i.e.,

$$\pi = aOJK - bOIK + cOIJ - dIJK, \tag{5}$$

is that their product $P\pi = 0$. Developing and suppressing the common factor $OIJK$, we have

$$ax + by + cz + d = 0, \tag{6}$$

the equation of the plane with the given coordinates.

The area of the triangle π is half the modulus of its bivector $\omega\pi = aJK + bKI + cIJ$, whence

$$\text{area } \pi = \tfrac{1}{2}\sqrt{a^2 + b^2 + c^2}. \tag{7}$$

Suppose we wish to find the distance δ of the point P from the plane π. We have

$$P\pi = \tfrac{1}{3}\delta\tfrac{1}{2} \text{mod}(\omega\pi)m^3,$$

where because of homogeneity we have written the factor m^3 (cubic metres). We observe that $OIJK = \tfrac{1}{6}m^3$, whence

$$\delta = \frac{P\pi}{\text{mod}(\omega\pi)OIJK},$$

and substituting the values of $P\pi$, $\omega\pi$, we have the formula sought.

As a last example, consider a form of second degree s, with coordinates l, m, n, p, q, r; i.e.,

$$s = lOI + mOJ + nOK + pJK + qKI + rIJ.$$

This formula says that s is the sum of the line $O(lI + mJ + nK)$ with origin the point O, and whose vector has coordinates l, m, n, and of the bivector with coordinates p, q, r. We wish to transform s into the sum of a line i and a bivector u which are mutually orthogonal

$$s = i + u.$$

We deduce that $\omega s = \omega i$, since $\omega u = 0$. Hence u, which has to be normal to $\omega i = \omega s$, will be of the form $u = x|\omega s$, where x is a (real) number.

We deduce that $s - x|\omega s = i$, and multiply this equation by itself, observing that $(|\omega s)(|\omega s) = 0$ and $ii = 0$. Solving, we have $ss - 2xs|\omega s = 0$, whence

$$x = \frac{ss}{2s|\omega s},$$

and finally

$$u = \frac{ss}{2s|\omega s}\,|\omega s.$$

We thus have the bivector u, called in mechanics 'the principal moment of the system of forces,' expressed by use of the form s alone. Introducing coordinates, we have

$$u = \frac{lp + mq + nr}{l^2 + m^2 + n^2}\,(lJK + mKI + nIJ).$$

The line i may be obtained by subtracting: $i = s - u$.

These elementary examples prove that the method of Grassmann by no means excludes ordinary analytic geometry, but on the contrary indicates very simple ways of finding formulas in every system of coordinates. Besides, in this way we have a separate geometrical significance for the numerator, the denominator, every factor, and every term of the formulas of analytic geometry.

12 INFINITESIMAL GEOMETRY

We say that the variable form of first degree S has as its limit the fixed form S_0 if, for an arbitrary triangle PQR, we have

$$\lim SPQR = S_0PQR.$$

Analogous definitions may be made for the forms of higher degree.

If a form $S(t)$ is a function of a numerical variable t, we let

$$\frac{dS(t)}{dt} = \lim_{h=0} \frac{S(t + h) - S(t)}{h},$$

where, in the second member, every sign introduced has previously been defined.

The common rules for derivatives hold for the sums and products of forms, and the symbols ω and | act as constant factors. We must be careful, however, not to invert the order of the factors.

The definitions of successive derivatives and of integrals are applicable here. The formula of Taylor holds in the form:

'If $S(t)$ is a geometrical form, a function of t, having first and second derivatives for $t = t_0$, we have

$$S(t_0 + h) = S(t_0) + h S'(t_0) + \frac{h^2}{2!} S''(t_0) + R,$$

where R is a form which is infinitesimal with h, of order higher than the second.'

There is no need to assume, for example, the continuity of $S''(t)$, as I have proved in my treatises.

The expression for the remainder, in the form of an integral, holds without any variation.

Lagrange's form for the remainder must be slightly modified, and since in my *Calcolo geometrico* (chapter 8, p. 133) I stated the result without giving a proof, it may be useful to give it here.

DEFINITION We say that a geometrical form S is intermediate among several others A, B, ... of the same degree if, for arbitrary P such that SP is a tetrahedron, we have that SP is intermediate among AP, BP, ...[2]

THEOREM Given the form $S(t)$, a function of the real variable t, having successive derivatives up to the nth in the interval from t to $t + h$, we have

$$S(t + h) = S(t) + h S'(t) + \ldots + \frac{h^{n-1}}{(n-1)!} S^{(n-1)}(t) + \frac{h^n}{n!} K,$$

where K is an intermediate geometrical form among the values of $S^{(n)}(t + \theta h)$, where θ varies between 0 and 1.

That is, while for numerical functions K is one of the values of the nth derivative, here, for geometrical forms, K is only intermediate among the values of the nth derivative.

In fact, the formula is true if $S(t)$ is a numerical function of t, since K is then one of the values of the nth derivative, and hence is intermediate among those which it may have. It is also true if $S(t)$ is a tetrahedron, or form of fourth degree, since tetrahedrons are measured by real numbers.

If $S(t)$ is a form of first degree, we multiply the given formula by an

2 [A tetrahedron is said to be intermediate among a set of tetrahedrons if its volume is intermediate among their volumes.]

arbitrary triangle PQR, and so reduce the problem to working with tetrahedrons. We deduce that $KPQR$ is intermediate among the values of $S^{(n)}(t)PQR$, whence K is intermediate among the values of $S^{(n)}(t)$.

If P is a point function of the real variable t, its derivatives are vectors. The tangent to the curve described by P is the straight line PP' and the osculating plane is the plane $PP'P''$. We suppose that this line and triangle are not zero; otherwise singularities are presented. These have been studied, for example, in my *Applicazioni geometriche*.

If the point P is a function of two variables t and u, the plane $P(dP/dt)(dP/du)$ is tangent to the surface described by P.

XVI

The most general question in the mathematical sciences (1896)*

This is the only article we include which was not listed in Cassina's Chronological Index of the Scientific Publications of Giuseppe Peano (Opere scelte, vol. I). We discover here to what extent Peano saw the Formulario project as a realization of the dream of Leibniz.

[What is the most general question that has been posed or resolved up to now in the whole of the mathematical sciences? *Lausbrachter.*]

Leibniz, who gave the fundamental rules of the infinitesimal calculus, and who was occupied with a geometrical analysis resembling the theory of vectors, also projected throughout his whole life, from his first work to his last letters, a 'universal characteristic' or 'general method by which all truths of the reason would be reduced to a kind of calculation.' He heaps praise on this project and says that 'no man, even though he be a prophet or a prince, can undertake anything of more benefit to mankind.' Thus, according to Leibniz, the most general question that has been posed is the construction of this characteristic.

After many studies, due to several authors, notably Boole and Schröder, the question has now been resolved. All the ideas of logic are expressed by a small number of signs on which we reason as on the signs $+$, $-$, $=$, ... of algebra.

I published in 1889, for the first time, a small book written entirely in the symbols of mathematical logic. This first work has been followed by several others, due to various authors. At the present time a society is publishing the *Formulaire de mathématiques*, which is to be a collection of all known propositions in various subjects in the mathematical sciences, with proofs and historical notes, all written with the symbols of logic. The first edition of the *Formulaire* appeared in 1895; at present a second edition is in the course of publication. Inquiries may be addressed to me.

---

* Reply to question no. 719, *L'intermédiaire des mathématiciens*, 3 (1896), 169 [90.5].

XVII

Studies in mathematical logic (1897)*

In 1897 Peano was very interested in logic per se, even though he afterwards said that his only interest in logic was its use in mathematics. In the following selection he is concerned with reducing the primitive ideas of logic to a minimum number (in this case, seven). The brief introduction is interesting for touching on Peano's contact with Frege. The article is notable for containing Peano's introduction of the symbol ∃.

Mathematical logic has interested me for many years. In the preliminary chapter to my *Calcolo geometrico* (1888) I explained in summary fashion the studies of Mr Schröder, in *Der Operationskreis des Logikkalküls* (1887), of Boole, and other authors. I pointed out there the identity of the calculus of classes, given by these authors, with the calculus of propositions, as found in the writings of Peirce, MacColl, et al.

Continuing this research, in my *Arithmetices principia, nova methodo exposita* (1889) I was fortunate enough to arrive at a complete analysis of the ideas of logic, reducing them to a quite limited number, which are expressed by the symbols ε, ⊃, =, ∩, ∪, ∼, Λ.

A result of this analysis was the construction of a graphic symbolism, or ideography, capable of representing all the ideas of logic, so that by introducing symbols to represent the ideas of the other sciences, we may express every theory symbolically.[1] For the first time, in that booklet, a complete theory was expressed in symbols; and it was precisely these which I used in order to distinguish what can from what cannot be defined (and what can from what cannot be proved) in arithmetic.

I used the same analytic method in other works: *I principii di geometria*,

---

* 'Studii di logica matematica,' *Atti Accad. sci. Torino*, 32 (1896–7), 565–83 [91].
1 With this ideography we are able at the present time to express the propositions of logic and of several mathematical theories, especially algebraic theories. To translate other theories into symbols, a complete analysis must be made of the ideas which occur in them, and these ideas must be reduced to symbols. Thus an ideography capable of representing, for example, all propositions of mathematics has only been partially constructed.

logicamente esposti (Turin, 1889), 'Démonstration de l'intégrabilité des équations différentielles' (*Math. Ann.*, 1890), 'Sur la définition de la limite d'une fonction' (*Amer. J. Math.*, 1894), etc. C. Burali-Forti developed these new theories in his *Logica matematica* (Milan, 1894), and he used them in numerous other works: 'Sulle classi derivate a destra e a sinistra' (*Atti Accad. Torino*, 1894), 'Sul limite delle classi variabili' (*Atti Accad. Torino*, 1895), 'Sur quelques propriétés des ensembles d'ensembles' (*Math. Ann.*, 1895), etc.

M. Pieri adopted the same method in analysing the foundations of *geometria di posizione* in a series of works published by this academy.

For several years a society has been publishing the *Formulaire de mathématiques*, whose 'Introduction' appeared in 1894. The first edition, begun in 1892, was completed in 1895. This publication is meant to contain, expressed in the symbols of logic, the theorems, definitions, and proofs of various mathematical theories. Collaborating in this work are Messrs Vailati, Castellano, Burali-Forti, Giudice, Vivanti, Betazzi, Fano, and others who send in additions and corrections. The second edition is now in the press, but many difficulties are delaying its publication.

This ideography, which is derived from studies in mathematical logic, is not just a conventional abbreviated writing, or tachygraphy, since our symbols do not represent words but ideas. For this reason one must write the same symbol wherever the same idea is found, whatever the expression used in ordinary language to represent it, and distinct symbols must be used where the same word is found when, because of its position, it represents a distinct idea. In this way we establish a one-to-one correspondence between ideas and symbols, a correspondence which is not found in our ordinary language. This ideography is based on the theorems of logic, discovered successively from Leibniz up to our day. The form of the symbols may be changed, i.e., the few signs to represent the fundamental ideas, but there cannot be two substantially different ideographies.

I have mentioned several works in which use was made of the ideography, which I shall call *mine*, despite the fact that two distinct ideographies cannot exist, for the following reason.

Mr G. Frege, a professor at the University of Jena, to whom many interesting works in mathematical logic are due, of which the first dates from 1879, has arrived in turn, and by a quite independent path,[2] in his

2 The works of Frege are independent of those of the numerous authors of mathematical logic. See, for example, *Symbolic Logic* of Venn (London, 1894, p. 493).
I am not now able to say whether the ideography of Frege is or is not complete, i.e. whether its symbolic propositions can be read independently of the accompanying text. Frege's formulas are, for me, much more difficult to read than those of other authors.

book *Grundgesetze der Arithmetik* (1893) at the expression in symbols of a series of propositions concerning the concept of number.

I published a brief mention of this book in the *Rivista di matematica*, 5 (1895), 122–8. Recently the same author published a note, 'Ueber die Begriffsschrift des Herrn Peano und meine eigene' (*Bericht. d. math. Classe d. Gesellschaft zu Leipzig*, 6 July 1896), where, although mentioning the Formulario and its Introduction, he expresses doubt that my ideography can serve to do more than express propositions, although in the works cited above its importance as a means of reasoning is evident.

I have to praise the impartiality of the judgments in the writing of Mr Frege, but even if we are in agreement on many points, our opinions still diverge on various questions, and this is because we attribute diverse meanings to several words and to several symbols.

Nevertheless, even if we regard this ideography as only a graphic symbolism capable of representing in a brief and precise form all the propositions of mathematics, its importance is evident. Further, this criterion of being able to use a symbolism as a language may be used to recognize whether it is complete or not.

Among the ideas of logic are numerous relations which may be expressed by theorems or formulas. I published a collection in my 'Formule di Logica matematica' (*Rivista di matematica*, 1891). This, with new formulas and numerous historical indications, due in great part to Dr Vailati, constitutes part I of the *Formulaire*, 1st ed. Numerous additions have been sent in by various correspondents, and a new edition is becoming more and more desirable.

Many of these formulas have the form of an equality, in which a sign is found in one member which is not found in the other member, or is found in it in a different position. Such equalities allow the expression of this sign by the use of the others, i.e., they can be assumed as the definition of this sign. Thus, with opportune definitions, one may reduce the ideas of logic to a smaller and smaller number of fundamental ideas, or primitive ideas, which must be expressed in ordinary language, or clarified with examples, but which cannot be expressed symbolically using others which are simpler. Still, this reduction of the ideas of logic to its fundamentals presents serious difficulties, and it is easier to recognize how many and which are the primitive ideas in arithmetic and geometry than in logic.

In this note we deal with the reduction of the ideas of logic to a minimum number. The significance of some symbols, explained in common language, having been admitted, all the other propositions will be written entirely in symbols, assuming nothing not explicitly stated and not leaving anything to be explained in words. In this way the formulas alone form an intelligible text, while the added clarifications and observations, in ordinary language, will facilitate its understanding.

PRIMITIVE IDEAS

The conventions which follow, and which we have to explain in ordinary language, represent primitive ideas:

1. The letters $a, b, ..., x, y, z$ indicate arbitrary objects that vary with the proposition.

2. Formulas are divided into parts using parentheses or dots. Thus the formulas

$$ab . c, \qquad a . bc, \qquad ab . cd, \qquad ab . cd : e . fg$$

are equivalent to

$$(ab)c, \qquad a(bc), \qquad (ab)(cd), \qquad [(ab)(cd)][e(fg)].$$

3. K signifies 'class.'

4. Let a be a K; $x \, \varepsilon \, a$ means 'x is an a.'

5. Let p and q be propositions containing variable letters $x, ..., z$. The formula

$$p \supset_{x,...,z} q$$

means 'whatever $x, ..., z$ are, as long as they satisfy the condition p, they will satisfy the condition q.' The indices to the sign $\supset$ are omitted when there is no danger of ambiguity.

6. The expression pq indicates the simultaneous affirmation of the propositions p and q.

The first convention is familiar to us from algebra and geometry. It was used by Aristotle in logic, but it is necessary to state it here, as well as the one about parentheses, since we wish to list all the conventions we shall use.

The ideas represented by our symbols are very simple ideas, and do not have exactly the value of their corresponding terms in ordinary language, which represent more complex ideas. Thus the sign ε may be read 'is a,' or 'est' in Latin, but represents the idea obtained from the term 'est' when abstraction is made from grammatical mood, tense, and person. Hence, there being no exact correspondence between the symbols and the terms of ordinary language, the exact value of the symbols is best learned, and more easily, from examples.

In order to give examples from arithmetic we shall use the symbols:

N   in place of 'positive integer'
Np   ”   'prime number'
N × a   ”   'multiple of a'

EXAMPLES

$7 \, \varepsilon \, \mathrm{Np}, \qquad 12 \, \varepsilon \, \mathrm{N} \times 4$

$a \, \varepsilon \, \mathrm{N} \, . \, \supset \, . \, a(a + 1)(a + 2) \, \varepsilon \, \mathrm{N} \times 6$

'If a is a number, then the product $a(a + 1)(a + 2)$ is a multiple of 6.'

$a \, \varepsilon \, \mathrm{Np} \, . \, \supset \, . \, (a - 1)! + 1 \, \varepsilon \, \mathrm{N} \times a$

'If a is a prime number, then the number expressed is a multiple of a' (Wilson).

In the above, the sign $\supset$ carries implicitly the letter a as index. These propositions consist of three parts, hypothesis, sign of deduction, and conclusion.

$x \, \varepsilon \, \mathrm{N} \, . \, x < 17 \, . \, \supset \, . \, x^2 - x + 17 \, \varepsilon \, \mathrm{Np}$

'Whatever integer x is, less than 17, the expression $x^2 - x + 17$ always represents a prime number' (Legendre).

$a \, \varepsilon \, \mathrm{Np} \, . \, b \, \varepsilon \, \mathrm{N} \, . \, b^2 \, \varepsilon \, \mathrm{N} \times a \, . \, \supset \, . \, b \, \varepsilon \, \mathrm{N} \times a$

'If the square of a number b is a multiple of a prime number a, b will also be a multiple of a' (Euclid).

Here the hypothesis is the simultaneous affirmation of several propositions. The sign $\supset$ carries as implicit indices the letters a and b.

We shall give an example in which the hypothesis itself contains the sign of deduction (the new signs of arithmetic which figure in it are immediately explained):

$a \, \varepsilon \, \mathrm{N} : x \, \varepsilon \, \mathrm{Np} \, . \, \supset_x \, . \, \mathrm{mp}(x, a) \, \varepsilon \, \mathrm{N}_0 \times 2 : \supset \, . \, a \, \varepsilon \, \mathrm{N}^2$

'If a is a number, and if whatever the prime number x the greatest power of x which divides a is an even number, including zero, then a is a perfect square.'

The correct use of the sign $\supset$ is intimately connected with that of the variable letters. Seeing that according to our conventions the letters $a, b, \ldots$ represent any variable entities whatever, in every proposition we must begin by saying what type of entity they represent. Hence the proposition

$a \times b = b \times a$

has no meaning for us, since it is incomplete. We must premise it with the meaning of the letters a and b, and write, for example,

$a \, \varepsilon \, \mathrm{N} \, . \, b \, \varepsilon \, \mathrm{N} \, . \, \supset \, . \, a \times b = b \times a.$

Instead of supposing a and b to be integers, we could suppose them to be fractions, irrationals, or imaginaries, and the conclusion would continue to hold, but it would be false if they were non-coplanar quaternions and it would be nonsense if a and b were entities for which multiplication had not been defined.

We say that a variable letter in a formula is apparent, or is an apparent variable, if its value in the formula is independent of the letter. Thus in $\int_a^b f(x)dx$ the letter x is apparent.

In every proposition, the letters connected to the sign $\supset$ are apparent, whether they are expressed or implied. Thus

$$x \,\varepsilon\, \text{Np} \,.\, \supset_x .\, \text{mp}(x, a) \,\varepsilon\, \text{N}_0 \times 2,$$

'whatever the prime number x, the greatest power of x contained in a is even,' is a proposition which expresses a condition on the letter a, and not on x, for which y could be substituted without changing the condition.

All letters which appear in a theorem are apparent, since the theorem expresses a truth which is independent of the letters used.

I have lingered a bit with the sign $\supset$ and its relative indices because there are divergences between the use Mr Frege and I make of my symbols. Indeed, in my usage the sign $\supset$ is essentially placed between propositions containing variable letters. Mr Frege, on the contrary, gives as examples of the sign $\supset$ the propositions

$$2^2 = 4 \,.\, \supset \,.\, 3 + 7 = 10,$$
$$2 > 3 \,.\, \supset \,.\, 7^2 = 0,$$

where the sign $\supset$ is found between propositions containing no variable letters.

Analogously, the example of Mr Frege

$$x > 2 \,.\, \supset \,.\, x^2 > 2$$

is not, according to me, complete, since whenever we introduce a letter x, we must begin by saying what it represents. We could complete it by writing, for example,

$$x \,\varepsilon\, \text{N} \,.\, x > 2 \,.\, \supset \,.\, x^2 > 2.$$

Mr Frege considers expressions of the form

$$(\Phi(x) \supset_x \Psi(x, y)) \supset_y \text{X}(y),$$

which are likewise not met with in the Formulario, because in a deduction it may happen that the hypothesis contains letters which are not in the conclusion, but it never happens that letters appear in the conclusion which are not in the hypothesis. Likewise, the example $(2 > 3) = \Lambda$ of Mr **Frege** is not found in the Formulario.

DEFINITIONS

Combining the primitive notations just introduced we may derive new ideas, which may then be defined in symbols. By a symbolic definition of a new sign x we mean the convention of representing by x a group of signs having an already known meaning, and we shall indicate this by writing

$x = a$ Def.

If what we define, x, contains variable letters, and if it is necessary to limit the meaning of these letters by using a hypothesis, the definition assumes the form

Hypothesis. $\supset . x = a$ Def.

The two signs $=$, Def., although written separately, must be considered as a single symbol, which is read 'is equal to, by definition' or 'we name.'

Let a be a K. We often have to write the proposition 'x and y are in a,' and shall make the convention of symbolically indicating this by $x, y \, \varepsilon \, a$. As this proposition has the value of two, $x \, \varepsilon \, a . y \, \varepsilon \, a$, we shall define:

1. $a \, \varepsilon \, \mathrm{K} . \supset : x, y \, \varepsilon \, a . = . x \, \varepsilon \, a . y \, \varepsilon \, a$ Def.

The common character of definitions, that of being an abbreviation clearly shows up in this example. Whoever does not wish to adopt this definition may write $x \, \varepsilon \, a . y \, \varepsilon \, a$ in place of $x, y \, \varepsilon \, a$, and the ideography which would result, whether adopting this definition or not, would not in substance be distinct at all. But this definition effectively gives a useful abbreviation, and so it is convenient to adopt it.

$x, y, z \, \varepsilon \, a$ means $x, y \, \varepsilon \, a . z \, \varepsilon \, a$, i.e., $x \, \varepsilon \, a . y \, \varepsilon \, a . z \, \varepsilon \, a$.

Let a and b be K. We shall write $a \supset b$ in place of 'every a is a b.' We may define this symbolically as follows:

2. $a, b \, \varepsilon \, \mathrm{K} . \supset \therefore a \supset b . = : x \, \varepsilon \, a . \supset_x . x \, \varepsilon \, b$.   Def.

In the formula $x \, \varepsilon \, a . \supset_x . x \, \varepsilon \, b$, 'if x is an a, then x is also a b,' the letter x which figures as index to the sign is an *apparent* letter, i.e., the value of the proposition does not depend on a. It thus expresses a relation between the letters a and b, which we may agree to indicate by $a \supset b$, where the apparent letter x is suppressed.

The sign ⊃ between classes may be read 'is contained in,' while between propositions it is read 'we deduce.' The fact that it may be read in several ways does not prove that it has several meanings, but only that ordinary language has several terms to represent the same idea. The term which would better correspond to the sign ⊃ in its various positions could possibly be 'hence,' 'ergo.'

EXAMPLE

$$N \times 6 \supset N \times 2$$

'every multiple of 6 is a multiple of 2,' or 'a multiple of 6, therefore multiple of 2,' is an application of definition 2. If one does not wish to make use of this definition, the same proposition may be written

$$x \varepsilon N \times 6 . \supset . x \varepsilon N \times 2$$

'If x is a multiple of 6, then x is a multiple of 2.' The sign ⊃ carries here the index x implicitly.

Let a be a K. Writing $x \varepsilon$ before it we have the proposition $x \varepsilon a$ containing the variable letter x.

Vice versa, if p_x is a proposition containing the variable letter x, by $\overline{x\varepsilon} \, p_x$ we mean the class of xs which satisfy the condition p_x. Thus, letting a be this class, i.e., letting

$$a = \overline{x\varepsilon} \, p_x,$$

the proposition p_x will be equivalent to $x \varepsilon a$

$$x \varepsilon a . = . p_x.$$

The sign — written over $x\varepsilon$ is the sign of inversion, used here because of the analogy with inversion of functions. The whole sign $\overline{x\varepsilon}$ may be read 'the x such that.' In the expression $\overline{x\varepsilon} \, p_x$, the letter x is apparent.

We wish to express this use of the expression $\overline{x\varepsilon} \, p_x$ in symbols. Since we have not formed symbols to say 'let p_x be a proposition containing the variable letter x,' it is necessary to replace the proposition p_x by the form $x \varepsilon a$, where a is a K. Hence, we write

3. $a \varepsilon K . \supset . \overline{x\varepsilon} \, (x \varepsilon a) = a$ Def.

'If a is a class, then the sign $\overline{x\varepsilon}$ set before the proposition $x \varepsilon a$ gives the class a again.' This definition effectively expresses the first member, not yet having a meaning, by means of the second. Apparently, however, it substitutes a long notation for a short. This happens because the proposition containing x is written in the form $x \varepsilon a$. But if it is written in another form, the preceding definition will carry a true simplification.

Let a and b be K. By $a \cap b$, or simply ab, we indicate the class of entities which are at the same time in a and b. The sign $\cap$ corresponds more or less to the conjunction *and*; the operation which it represents is called *logical multiplication*. This operation may be defined by

4. $a, b \, \varepsilon \, \text{K} . \supset . \, ab = \overline{x\varepsilon} \, (x \, \varepsilon \, a . x \, \varepsilon \, b)$ Def.

'If a and b are classes, by ab we mean the set of x which satisfy the condition $x \, \varepsilon \, a . x \, \varepsilon \, b$.'

Operating with the sign $x \, \varepsilon$ on both the members of this equality we have

$a, b \, \varepsilon \, \text{K} . \supset : x \, \varepsilon \, ab . = x \, \varepsilon \, a . x \, \varepsilon \, b$

'To say that x is an ab is equivalent to saying that x is an a and x is a b.' This equality, however, could not serve as a definition of the symbol ab, but only of the whole expression $x \, \varepsilon \, ab$.

Thus logical multiplication of classes is defined by means of simultaneous affirmation, or logical multiplication of propositions, which was assumed as a primitive idea, and by means of the sign $\overline{x\varepsilon}$ which was already defined (Def. 2). But I have not succeeded in expressing the sign ab without making use of Def. 3.

EXAMPLE

$\text{Np} \cap (4\text{N} + 1) \supset \text{N}^2 + \text{N}^2$

'Every prime number of the form $4x + 1$, where x is an N, is the sum of two squares.' Not wishing to make use of the definitions introduced but only of the primitive ideas, we may write this proposition

$x \, \varepsilon \, \text{Np} . x \, \varepsilon \, 4\text{N} + 1 . \supset . x \, \varepsilon \, \text{N}^2 + \text{N}^2.$

We shall make the following definition:

5. $a, b \, \varepsilon \, \text{K} . \supset : a = b . = . a \supset b . b \supset a$ Def.

'Let a and b be classes. We shall say that $a = b$ if every a is a b and every b is an a.' In this definition is found, in the first member, the sign $=$ between classes, a sign we wish to define, and in the second member an expression not containing this sign. The two members are connected with the sign $=$, but this must be considered as united with the sign Def., so that the whole sign $=$ Def. must be considered as a single sign. Thus the vicious circle of defining the sign $=$ by using the same sign is only apparent.

The following propositions are worth mentioning:

$a, b, c \, \varepsilon \, \text{K} . \supset . \, aa = a,$

$ab = ba,$

$a(bc) = (ab)c;$

these have already been stated in words by Leibniz (*Opera philosophica* [1840], p. 98) and in symbols by Boole (*An Investigation of the Laws of Thought* [1854], pp. 29, 31), except for the meaning of the letters, which still had to be expressed in ordinary language.

EXAMPLE

$$(N \times 2) \cap (N \times 3) = (N \times 6)$$

The sign Λ, in a discussion of classes, indicates the null class, i.e., the class which contains no individuals. It may be defined as follows:

6. $a \, \varepsilon \, K . \supset \, \therefore \, a = \Lambda . = \, : b \, \varepsilon \, K . \supset_b . a \supset b$ Def.

'Let a be a class. We shall say that the class a is null if, however we choose the class b, the class a is contained in b.'

The proposition $b \, \varepsilon \, K . \supset_b . a \supset b$ contains the apparent letter b, and hence is a condition only on a, so that we may agree to indicate it by the expression $a = \Lambda$, where only the letter a appears.

Note that what has been defined is just the proposition $a = \Lambda$ and hence, for now, the complex of signs $= \Lambda$ must be considered as a single sign. The notation is advantageous, however, since this condition $a = \Lambda$ acts as an equality. That is to say, we may prove the propositions

$$a, b \, \varepsilon \, K . a = \Lambda . b = \Lambda . \supset . a = b,$$
$$\text{''} \quad . a = b . b = \Lambda . \supset . a = \Lambda.$$

The sign Λ remains for the moment undefined, i.e., we may not yet form an equality whose first member is Λ and whose second is a group of known signs.

Analogously to the sign Λ we may introduce the sign V (all):

7. $a \, \varepsilon \, K . \supset \, \therefore \, a = V . = \, : b \, \varepsilon \, K . \supset_b . b \supset a.$

The sign V has no practical use, however, and is found nowhere in the Formulario.

EXAMPLE

$$N^3 \cap (N^3 + N^3) = \Lambda$$

'Cubic numbers which are sums of cubes do not exist.' If we wish to express this proposition without making use of the definitions, but only of the primitive ideas, it becomes

$$x \, \varepsilon \, N^3 . x \, \varepsilon \, N^3 + N^3 . a \, \varepsilon \, K . \supset . x \, \varepsilon \, a.$$

Let a and b be classes. The expression $a \cup b$ indicates the smallest class containing a and b. The sign $\cup$ is read *or*; the operation indicated by this sign is called *logical addition*.

This sign may be defined by the preceding symbols, as follows:

8.　$a, b \, \varepsilon \, \mathrm{K} . \supset . \, a \cup b = \overline{x\varepsilon} \, (c \, \varepsilon \, \mathrm{K} . a \supset c . b \supset c . \supset_c . \, x \, \varepsilon \, c)$　　　　Def.

'If a and b have the meaning given, $a \cup b$ indicates the set of individuals pertaining to every class c which contains the two classes a and b.'

We have

$a, b, c \, \varepsilon \, \mathrm{K} . a \supset c . b \supset c . \supset . a \cup b \supset c$　　　　(Leibniz, p. 96)

This expresses the distributive property of logical multiplication with respect to addition, a property stated by Lambert in 1781.

EXAMPLE

$\mathrm{Np} \cap (3 + \mathrm{N}) \supset (6\mathrm{N} - 1) \cup (6\mathrm{N} + 1)$

This proposition is expressed without making use of definition 8 in this way:

$x \, \varepsilon \, \mathrm{Np} . x > 3 . a \, \varepsilon \, \mathrm{K} . 6\mathrm{N} - 1 \supset a . 6\mathrm{N} + 1 \supset a . \supset . x \, \varepsilon \, a.$

If a is a class, by $\sim a$ we mean the class of the non-a, which can be defined as follows:

9.　$a \, \varepsilon \, \mathrm{K} . \supset . \sim a = \overline{x\varepsilon} \, (b \, \varepsilon \, \mathrm{K} . a \cup b = \mathrm{V} . \supset_b . \, x \, \varepsilon \, b)$　　　　Def.

'By $\sim a$ we mean the set of x pertaining to every class b which with a gives the whole as sum.'

Thus negation is expressed by means of the signs $\cup$ and V. Among the many identities which we have, we shall mention two:

$$a, b \, \varepsilon \, \mathrm{K} . \supset . \sim(a \cup b) = (\sim a) \cap (\sim b)$$
$$\sim(a \cap b) = (\sim a) \cup (\sim b)$$

expressed in symbols (except for the significance of the letters) by De Morgan in 1858.

From the first we derive

$a, b \, \varepsilon \, \mathrm{K} . \supset . a \cup b = \sim [(\sim a) \cap (\sim b)],$

which could serve as a definition of the sign $\cup$ using the signs $\sim$ and $\cap$ (as was done in the Formulario), but the present selection leads to a further reduction.

The signs $\supset$ and $\cap$ can be found between propositions, or between classes, and the meaning of the second is deduced from the first, by means of Definitions 2 and 4. The signs $=$, Λ, $\cup$, $\sim$, defined for classes, occur also in propositions, and will be defined as follows:

10. $a, b \,\varepsilon\, K . \supset \therefore x \,\varepsilon\, a . =_x . x \,\varepsilon\, b : = \, : x \,\varepsilon\, a . \supset_x .$
    $x \,\varepsilon\, b : x \,\varepsilon\, b . \supset_x . x \,\varepsilon\, a$ Def.

or   ,,   ,, $= . a = b$

'We shall say that the two conditional propositions in x, $x \,\varepsilon\, a$ and $x \,\varepsilon\, b$, are equivalent with respect to x, if the second is deduced from the first and vice versa or, what is the same thing, if the classes a and b are equal.'

The signs $\supset$ and $=$ can, and often do, occur in another position. We make the following definition:

11. $a, b, c \,\varepsilon\, K . \supset \,:: x \,\varepsilon\, a . \supset_x : x \,\varepsilon\, b . \supset . x \,\varepsilon\, c \therefore = \, :$
    $x \,\varepsilon\, a . x \,\varepsilon\, b . \supset_x . x \,\varepsilon\, c$

or   ,,   ,, $= . ab \supset c$ Def.

'If a, b, c are classes, we shall say that from $x \,\varepsilon\, a$ is deduced with respect to x that from $x \,\varepsilon\, b$ is deduced $x \,\varepsilon\, c$, if and only if from $x \,\varepsilon\, a$ and from $x \,\varepsilon\, b$ is deduced $x \,\varepsilon\, c$, i.e., if the class ab is contained in c.'

12. $a, b, c \,\varepsilon\, K . \supset \,:: x \,\varepsilon\, a . \supset_x : x \,\varepsilon\, b . = . x \,\varepsilon\, c \therefore = .$
    $ab \supset c . ac \supset b$ Def.

'We shall say that if $x \,\varepsilon\, a$, then the condition $x \,\varepsilon\, b$ is equivalent to $x \,\varepsilon\, c$ if and only if on the assumption that $x \,\varepsilon\, a$, from $x \,\varepsilon\, b$ is deduced $x \,\varepsilon\, c$ and vice versa, that is, when $ab \supset c$, and $ac \supset b$.'

13. $a \,\varepsilon\, K . \supset \therefore x \,\varepsilon\, a . =_x \Lambda : = . a = \Lambda$ Def.

'If a is a class, we shall say that the proposition $x \,\varepsilon\, a$ is absurd with respect to the variable x (written as above) if and only if the class a is null.'

14. $a, b \,\varepsilon\, K . \supset \, : x \,\varepsilon\, a . \cup . x \,\varepsilon\, b . = . x \,\varepsilon\, a \cup b$ Def.

'We shall write $x \,\varepsilon\, a . \cup . x \,\varepsilon\, b$ (which we shall read "x is an a, or x is a b") instead of x is an a or a b.'

15. $a \,\varepsilon\, K . \supset \, : \sim(x \,\varepsilon\, a) . = . x \,\varepsilon\, \sim a,$ Def.

which expresses the negation of a proposition by means of the negation of a class.

In these definitions the first member is more complicated than the second, which results from our having expressed the proposition on which we operate in the form $x \,\varepsilon\, a$.

In order to suppress parentheses, often we write the sign $\sim$ before the sign of the relation, i.e.,

16. $a \,\varepsilon\, K . \supset : x \sim \varepsilon\, a . = . \sim (x \,\varepsilon\, a)$,　　　　　　　　Def.

17. $x \sim = y . = . \sim (x = y)$.　　　　　　　　Def.

So that several of the preceding definitions may receive all the generalization necessary in our formulas, it is necessary to introduce the concept of a pair.

The sign $(x; y)$ indicates the pair formed by the objects x and y.

This pair is considered as a new object. In the Formulario, instead of $(x; y)$ is written simply (x, y), there being, in practice, no ambiguity with Definition 1.

The idea of a pair is fundamental, i.e., we do not know how to express it using the preceding symbols.[3] We can, however, define the equality of two pairs:

18. $(x; y) = (a; b) . = . x = a . y = b$　　　　　　　　Def.

'The pair $(x; y)$ is said to be equal to $(a; b)$ when their elements are equal in the same order.'

With the idea of a pair we can express several important rules of reasoning, which have already been explained in ordinary language in my works, but which we may now express completely in symbols. One of them is:

$$a, b, c \,\varepsilon\, K : x \,\varepsilon\, a . (x; y) \,\varepsilon\, b . \supset_{x,y} . (x; y) \,\varepsilon\, c : \supset \therefore$$
$$x \,\varepsilon\, a . \supset_x : (x; y) \,\varepsilon\, b . \supset_y . (x; y) \,\varepsilon\, c$$

'Let a, b, c be classes. Let it be supposed that whatever x and y are, if x pertains to the class a, and if the pair $(x; y)$ pertains to the class b, then the pair pertains to the class c. Then, whatever x is, provided that it is an a, we deduce that whatever y is, provided that the pair $(x; y)$ satisfies the condition b, the pair $(x; y)$ will satisfy the condition c.'

This rule is called 'exportation' or 'separation of the hypotheses.' The inverse proposition also holds.

EXAMPLE

$$a \,\varepsilon\, N . b \,\varepsilon\, N \times a . c \,\varepsilon\, N \times b . \supset . c \,\varepsilon\, N \times a.$$

Here the sign $\supset$ carries implicitly the indices a, b, c. Exporting the hypotheses relative to a and b, we have

$$a \,\varepsilon\, N . b \,\varepsilon\, N \times a . \supset : c \,\varepsilon\, N \times b . \supset_c . c \,\varepsilon\, N \times a.$$

3 [This was done by Norbert Wiener in 1914: 'A Simplification of the Logic of Relations,' *Proc. Camb. Phil. Soc.*, 17 (1912–14), 387–90.]

The first sign $\supset$ carries implicitly the indices a and b, and the second the written index c. By Definition 2 we may also write

$$a \,\varepsilon\, N . b \,\varepsilon\, N \times a . \supset . N \times b \supset N \times a.$$

The triplet $(x; y; z)$ may be considered as a pair formed of $(x; y)$ and z.

Definitions 10–15 express operations on propositions of the form $x \,\varepsilon\, a$, i.e., propositions containing only the variable letter x. We may assume, however, that x represents a pair, or a triplet, or any system whatever of letters. Hence, with the concept of pair, these definitions express operations on any conditional propositions whatever.

The proposition $a \sim\, = \Lambda$, where a is a class, thus signifies 'some a exist.' Since this relation occurs rather often, some workers in this field hold it useful to indicate it by a single notation, instead of the group $\sim\, = \Lambda$. The following definition may be made:

19. $\quad a \,\varepsilon\, K . \supset : \exists\, a . = . a \sim\, = \Lambda$ \hfill Def.

EXAMPLE

$$\exists\, N^2 \cap (N^2 + N^2)$$

'There exist squares which are the sum of two squares.'

The sign $=$ has already been defined between two classes, between two propositions, between two pairs; and in the mathematical sciences it is defined anew each time it is found between entities newly introduced.

We may give the general definition

20. $\quad x = y . = : a \,\varepsilon\, K . x \,\varepsilon\, a . \supset_a . y \,\varepsilon\, a$ \hfill Def.

'We shall say that the entity x is equal to the entity y when every class a containing x also contains y.'

But it remains to be seen how far the various particular definitions enter into this one.

If x is any object whatever, by ιx we mean the class formed by this object alone:

21. $\quad \iota x = \bar{y}\varepsilon\, (y = x)$ \hfill Def.

'By ιx we mean the set of y which satisfy the condition $y = x$.'

We have the equalities

$$a \,\varepsilon\, K . \supset : x \,\varepsilon\, a . = . \iota x \supset a,$$
$$" \qquad x \,\varepsilon \sim a . = . \iota x \cap a = \Lambda,$$
$$" \qquad x, y \,\varepsilon\, a . = . \iota x \cup \iota y \supset a,$$

which express the propositions $x \,\varepsilon\, a$ and $x \,\varepsilon\, \sim a$, using others in which the signs ε, $\sim$ are not found.

Vice versa, let a be a class containing only one individual, i.e., let there exist some a, and however we take two individuals x and y of a, let them always be equal. This individual we shall indicate by $\bar{\imath}a$. Thus

22. $a \,\varepsilon\, K \,.\, \exists\, a : x, y \,\varepsilon\, a \,.\, \supset_{x,y} .\, x = y : \supset \,:\, x = \bar{\imath}a \,.\, = .\, a = \imath x.$ Def.

In fact this definition gives the significance of the whole formula $x = \bar{\imath}a$, and not of the single group $\bar{\imath}a$. But every proposition containing $\bar{\imath}a$ is reducible to the form $\bar{\imath}a \,\varepsilon\, b$, where b is a class; and this to a $\supset b$, where the sign $\bar{\imath}$ does not appear. Be this as it may, we do not succeed in forming an equality whose first member is $\bar{\imath}a$ and the second a group of known signs.

EXAMPLE

$a, b \,\varepsilon\, N \,.\, a < b \,.\, \supset\, .\, b - a = \bar{\imath}N \,\cap\, \overline{x\varepsilon}\,(a + x = b)$

'If a and b are some N, and $a < b$, then $b - a$ indicates that number x which added to a gives b.'

Definition 6 gives us the significance of the whole symbol $= \Lambda$; the symbol Λ can be defined:

$\Lambda = \bar{\imath}K \,\cap\, \overline{a\varepsilon}\,(b \,\varepsilon\, K \,.\, \supset_{b} .\, a \supset b)$

or $\quad \Lambda = \bar{\imath}\,\overline{x\varepsilon}\,(a \,\varepsilon\, K \,.\, \supset_{a} .\, a \sim a = x),$

'the null class is the common value of the expression $a \sim a$, whatever the class a is.'

We also have

$a \,\varepsilon\, K \,.\, \supset\, .\, \sim a = \bar{\imath}K \,\cap\, \overline{x\varepsilon}\,[a \,\cap\, x = \Lambda \,.\, a \,\cup\, x = V],$

'if a is a class, $\sim a$ indicates that class x such that multiplying by a gives the null class, and summing with a gives the whole.' This is the definition of the negation of a given by Mr Schröder in his *Algebra der Logik* of 1891, p. 32, partly in symbols and partly in words.

We shall also give the definition of correspondence (f):[4]

23. $a, b \,\varepsilon\, K \,.\, \supset\, \therefore\, f \,\varepsilon\, b\, f\, a \,.\, = \,:\, x \,\varepsilon\, a \,.\, \supset_{x} .\, fx \,\varepsilon\, b$

'Let a and b be classes. We shall say that f is a correspondence between the a and the b if writing the sign f before any individual whatever of a we obtain a b.'

4 [I have kept Peano's terminology for the next three definitions, but we may also say that Definition 23 defines '(single-valued) mapping of a into b,' Definition 24 defines 'one-to-one mapping of a into b,' and Definition 25 defines 'one-to-one mapping of a onto b.']

We may define the similarity correspondence:

24. $a, b \, \varepsilon \, \mathrm{K} \, . \, \supset \, . \, . \, f \varepsilon \, (b \, \mathrm{f} \, a) \mathrm{Sim} \, . \, = \, : f \varepsilon \, b \, \mathrm{f} \, a \, : x, y \, \varepsilon \, a \, .$

   $x \sim \, = y \, . \, \supset_{x,y} . \, fx \sim \, = fy$

and the reciprocal correspondence:

25. $a, b \, \varepsilon \, \mathrm{K} \, . \, \supset \, . \, . \, f \varepsilon \, (b \, \mathrm{f} \, a) \mathrm{rcp} \, . \, = \, : f \varepsilon \, (b \, \mathrm{f} \, a) \mathrm{Sim} \, : y \, \varepsilon \, b \, .$

   $\supset_y . \, \exists \, \overline{x \varepsilon} \, (x \, \varepsilon \, a \, . \, fx = y)$

and finally *number*.

26. $a, b \, \varepsilon \, \mathrm{K} \, . \, \supset \, : \mathrm{Num} \, a = \mathrm{Num} \, b \, . \, = \, . \, \exists \, (b \, \mathrm{f} \, a) \mathrm{rcp}$

'We shall say that the number of a is equal to the number of b if there exists a reciprocal correspondence between the a and the b.'

Now, having admitted several symbols explained in ordinary language and representative ideas, taken as primitive, we are able to give the symbolic definition of all the signs which are encountered in mathematical logic. I believe, however, that there remains much to be done in this field. For example, we may seek to reduce the number of ideas held to be primitive further, or try other paths, assuming as primitive ideas another set of ideas, so as to obtain a certain simplicity.

The numerous logical equalities which are known, and those which may be found, allow more than one way of classifying the symbols of logic. Thus the classification of logical propositions into primitive and derived would take a longer study. Let it suffice here to call to the attention of scholars these highly useful and interesting subjects.

XVIII
Supplement to
'On the Cantor-Bernstein Theorem' (1906)*

In 1906 Peano published in the Rendiconti del Circolo matematico di Palermo *a brief note on the Cantor-Bernstein theorem: Given two cardinal numbers x and y, if $x \geqslant y$ and $x \leqslant y$, then $x = y$. He was replying to some remarks of Poincaré on the same subject. Later that year he reprinted his note in the* Rivista di matematica *and added a slightly longer supplementary note. The following selection gives only this supplementary note. It has, in fact, little connection with the previous note. In it Peano discusses several antinomies and gives his own solution of Richard's antinomy. He also discusses the Postulate of Zermelo, pointing out that he, Peano, had already discussed this postulate in 1890 (i.e., fourteen years before Zermelo's statement of the postulate).*

The *Revue de métaphysique et de morale* has published two new papers: L. Couturat, 'Pour la logistique,' 14 (1906), 208–50; H. Poincaré, 'Les mathématiques et la logique,' 14 (1906), 294–317. Several other authors have joined the discussion, which continues to become larger and more interesting. The prime subjects being discussed are the several contradictions, or antinomies, which have now been found in some topics in mathematics.

In every age there has been some antinomy or other, which after some discussion received a solution. The solution of an antinomy is the indication of the point where there has been an error in the reasoning. Among the Greek philosophers there was a celebrated contradiction called 'Achilles and the tortoise.'[1] It comes from the equality

$$1 = 1/2 + 1/4 + 1/8 + ...,$$

where 1 is equal to a quantity which, although it varies, is always less than 1.

---

* 'Super theorema de Cantor-Bernstein et additione,' *Rivista de matematica*, 8 (1902–6), 136–57 [133′].
1 B. Russell, *The Principles of Mathematics* (Cambridge, 1903), p. 358.

In the seventeenth century one subject of discussion was the series

$$1 - 1 + 1 - 1 + ...,$$

which was equal to 1, 0, 1/2, etc., according to the law adopted in considering limits. An exact definition of 'limit' eliminates the antinomies cited.

The principal antinomy being discussed today was found in the theory of transfinite numbers by C. Burali-Forti, *Rend. Circ. mat. Palermo*, 11 (1897), 154–64. B. Russell (*The Principles of Mathematics* [Cambridge, 1903], p. 323) studied the proof of Burali-Forti and constructed new antinomies, similar to the preceding one, but simpler. P.E.B. Jourdain, in the present issue (*R.d.M.*, 8 [1906], 121–36) presents and amply discusses this question, and I have nothing to add.

Zermelo (*Math. Ann.*, 59 [1904], 514–16) proposed a new principle which has been allowed by some authors, denied by others, and discussed together with the antinomy of Burali-Forti by: König, *Math. Ann.*, 60 (1905), 177–80; Schönflies, *ibid.*, 181–6; Bernstein, *ibid.*, 187–93; Borel, *ibid.*, 194–5; Bernstein, *ibid.*, 463–4; Jourdain, *ibid.*, 465–70; König, *Math. Ann.*, 61 (1905), 156–60; and Hadamard, Borel, Baire, Lebesque in 'Cinq lettres sur la théorie des ensembles,' *Bulletin de la Société mathématique de France*, 33 (1905), 261–73.

J. Richard (*Revue générale des sciences pures et appliquées*, 16 [1905], 541) presented and explained a new antinomy.

Among the latest publications I may cite B. Russell, 'The Theory of Implication,' *American Journal of Mathematics*, 28 (1906), 159–202, and König, 'Sur la théorie des ensembles,' *C.R. Acad. Sci. Paris*, 143 (1906), 110–12.

Every antinomy, whether ancient or modern, depends on a consideration of the 'infinite.' The advice, given by some authors, not to consider the infinite is prudent, but does not resolve the problem, for the infinite is in the nature of several questions and 'naturam expelles furca, tamen usque recurret.'[2]

1 THE PRINCIPLE OF ZERMELO

Zermelo, in proving a proposition about well-ordered classes, adopted and stated the principle 'that even for an infinite totality of sets there are always mappings that associate with every set one of its elements ... This logical principle cannot, to be sure, be reduced to a still simpler one, but it is applied without hesitation everywhere in mathematical deduction.'

This principle means that we may arbitrarily choose an infinite number

---

2 [Horace, Ep. i. 10. 24. 'Chase nature out with a pitchfork, it keeps hurrying back.']

of elements. This assumption, which occurs in several books, was already considered by me in the year 1890, in *Math. Ann.*, 37, p. 210: 'one may not apply an infinite number of times an *arbitrary law* according to which to a class *a* is made to correspond an individual of that class ...'

Indeed, the form of argument 'if I arbitrarily choose an element x of class *a*, then proposition p (which does not contain x) follows' is reducible to the form

$$\exists\, a \qquad\qquad (1)$$
$$x\,\varepsilon\,a\,.\,\supset\,.\,p \quad (2)$$
$$(1)\,.\,(2)\,.\,\supset\,.\,p$$

'If there exists an *a*, and if from $x\,\varepsilon\,a$ follows proposition p, then proposition p may be affirmed.'

This is the form of argument called 'elimination of x' in *Formulario*, v, p. 12, Prop. 3.1. It is reduced to the rule of 'importation' in *Formulario*, II, Prop. 74, 310, 331, 405.

The assumption of two successive arbitrary elements has the form:

$$\exists\, a \qquad\qquad\qquad (1)$$
$$x\,\varepsilon\,a\,.\,\supset\,.\,\exists\, b \qquad (2)\ b\ \text{is a class which may contain}\ x$$
$$x\,\varepsilon\,a\,.\,y\,\varepsilon\,b\,.\,\supset\,.\,p \qquad (3)\ p\ \text{is a proposition independent of}\ x\ \text{and}\ y$$
$$(1)\,.\,(2)\,.\,(3)\,.\,\supset\,.\,p$$

For example, to prove the theorem

$$u\,\varepsilon\,\text{Cls'q}\,.\,\supset\,.\,\lambda\lambda u \supset \lambda u$$

'If u is a class of real numbers, then the limit class of the limit of u is contained in the limit class of u' (*Formulario*, v, p. 139, Prop. 1.2), I first affirm that

$$z\,\varepsilon\,\lambda\lambda u\,.\,h\,\varepsilon\,Q\,.\,\supset\,.\,\exists\,\lambda u \cap x\,\mathbf{\vartheta}\,[\text{mod}(x-z) < h/2] \qquad\qquad (1)$$

'If z is an element of $\lambda\lambda u$, and h is a positive real number, then there exists an individual in the class λu and in the class of the x whose distance from z is less than $h/2$.'

The truth of this pertains to mathematics, and not to logic.

$$\text{Hp}(1)\,.\,x\,\varepsilon\,\lambda u\,.\,\text{mod}(x-z) < h/2\,.\,\supset\,.$$
$$\exists\,u \cap y\,\mathbf{\vartheta}\,[\text{mod}(y-x) < h/2] \qquad\qquad (2)$$

'If z and h have the same meaning as in the hypothesis of (1), and if x is a λu, distant from z by less than $h/2$, then there exists an element in the class u and in the class of the y whose distance from x is less than $h/2$.'

$$\text{Hp}(2)\,.\,y\,\varepsilon\,u\,.\,\text{mod}(y-x) < h/2\,.\,\supset\,.\,\text{mod}(y-z) < h \qquad (2')$$

'If z, h, x have values as above, and if y is a u, distant from x by less than $h/2$, then the distance of y from z is less than h.'

I eliminate y in the thesis by the rule of *Formulario*, v, p. 12, Prop. 3.1:

$$\text{Hp}(2') . \supset . \exists\, u \cap y \,\textit{з}\; [\text{mod}(y - z) < h] \tag{3}$$

Note that the letter y in the thesis is apparent, and can be substituted for by any other letter a, ...; there is no letter y in the hypothesis.

The hypothesis of (3) contains the letters x and y. From (1), (2), (3) follows, by elimination of x and y,

$$z\,\varepsilon\,\lambda\lambda y . h\,\varepsilon\,\text{Q} . \supset . \exists\, u \cap y \,\textit{з}\; [\text{mod}(y - z) < h], \tag{4}$$

whence, by a rule of mathematics,

$$z\,\varepsilon\,\lambda\lambda u . \supset . z\,\varepsilon\,\lambda u,$$

and, after operating with $z\,\textit{з}$,

$$\lambda\lambda u \supset \lambda u.$$

The assumption of two arbitrary elements x and y leads to an argument with three hypotheses (1), (2), (3), and a thesis (4). In general the assumption of n successive arbitrary elements leads to an argument which consists of $n + 2$ propositions. Therefore we may not suppose $n = \infty$, that is, we cannot construct an argument with an infinite number of propositions.

In the Formulario, the word 'syllogism' indicates one well-defined form of argument:

$$a \supset b . b \supset c . \supset . a \supset c$$

'If from a follows b, and from b follows c, then from a follows c,' or 'if every a is a b, and every b is a c, then every a is a c.'

This corresponds to the 'syllogism in Barbara' of the scholastics.

In treatises of ordinary logic the word 'syllogism' has a wider meaning, not always definite.

The form of argument called 'elimination' in the Formulario consists of three propositions, and has three terms: a major a, a middle x, and a minor p. Therefore it may be called a syllogism, in the wider sense. Then every form of argument in the Formulario, Part I, is a syllogism, or a sorites (a chain of syllogisms). Thus the proposition of the scholastics, that every form of argument is reducible to a syllogism, makes sense. The assumption of one arbitrary element, or of several, if finite in number, is a syllogism, or a sorites with a finite number of premises.

Mr Poincaré replied to the question of Zermelo (p. 313): 'The axioms in question will only be propositions that some will admit ... and others will doubt ... There is nonetheless one point on which everyone agrees. The axiom is evident for finite classes.'

The question of evidence is subjective, and is not a part of mathematics.

We may affirm: 'The assumption of a finite number of arbitrary elements is a form of argument which is reducible to a system of syllogisms (in the broad sense). If any author gives a proof that is not reduced to syllogisms, and if he affirms that he has reduced the proof to syllogisms, he lies.'

In several cases the principle of Zermelo is reducible to syllogisms. See, for example, the proof in the cited paper in the *Mathematische Annalen*, vol. 37, where a function f such that $fu \, \varepsilon \, u$, if u is a closed non-empty class of complex numbers of order n, is constructed. Some propositions, which are proved by several authors with an (implicit) application of the principle of Zermelo, are proved in the Formulario without it.

That is, of several propositions some authors have given incomplete proofs, which can be completed.

In some cases we do not know how to eliminate the postulate of Zermelo. Then these proofs are not reduced to the ordinary forms of argument, and the proofs are not valid according to the ordinary meaning of the word 'proof.'

For example, in Borel, *Théorie des fonctions* (1898), p. 13, the proposition

$$a \, \varepsilon \, \mathrm{Cls} \, . \, \mathrm{Num} \, a \, \varepsilon \, \mathrm{infn} \, . \, \supset \, . \, \exists \, \mathrm{Cls}'a \, \cap \, b \, \mathbf{3} \, (\mathrm{Num} \, b = \aleph_0),$$

'If a is a class of objects infinite in number, then there exists in a numerable classes, that is, that have as their number aleph zero $=$ Num N_0,' is proved with the postulate of Zermelo, and the elimination of it is not easy. Therefore the proposition has not been proved.

Now, must we give an opinion as to whether the proposition is true or false? Our opinion does not matter. The preceding theorem is similar to the theorem of Goldbach

$$2(N_1 + 1) \supset Np + Np,$$

'Every even number 4, 6, 8, ... is the sum of two prime numbers,' which lacks a satisfactory proof.

2 CANTOR'S THEOREM

In order to understand the 'antinomy' considered by Richard, we present Cantor's theorem, expressed in symbols in the *Formulario*, v, p. 138, Prop. 11.1:

$$f \varepsilon \, \Theta \, \mathrm{f} N_0 \, . \, \supset \, . \, \exists \, \Theta - f \, `N_0$$

'If f is a sequence of real numbers in the interval from 0 to 1, then there exists a number in this interval which does not belong to the sequence f,' i.e., 'A denumerable class in the interval cannot constitute the whole interval.'

The proof given in the Formulario is:

$$\sum [10^{-n} \text{rest}(\text{Cfr}_{-n}fn + 5, 10)|n, N_1] \varepsilon \Theta - f \text{'}N_0$$

Consider the integer n; then fn is the nth number in the sequence and its nth decimal digit is indicated in the Formulario, p. 102, by $\text{Cfr}_{-n}fn$.

We change this digit. For example, we add 5, and then subtract 10 if the sum is greater than or equal to 10, giving

$$\text{rest}(\text{Cfr}_{-n}fn + 5, 10).$$

Then we form the number which has this digit as its nth decimal digit. It does not belong to the sequence. In fact, for every value of n it is different from fn, for its nth digit is different from the corresponding digit of fn, and we never have the case of the form 0.999... = 1.000..., where two decimal fractions of different forms have the same value.

For brevity, we let:

$$x \varepsilon 0 \ldots 9 \, . \, \supset \, . \text{ anti } x = \text{rest}(x + 5, 10) \qquad \text{Def.}$$

i.e., if

$$x = 0, 1, 2, 3, 4, 5, 6, 7, 8, 9,$$
$$\text{anti } x = 5, 6, 7, 8, 9, 0, 1, 2, 3, 4.$$

The number considered has the form

$$\sum [10^{-n} \text{ anti } \text{Cfr}_{-n}fn|n, N_1].$$

Instead of the antidigits of the digits of the given numbers, we could adopt some other law. But we may not take, for every value of n, an arbitrary digit from among the digits differing from $\text{Cfr}_{-n}fn$ without introducing the postulate of Zermelo.

3 RICHARD'S ANTINOMY

Richard, in the paper cited, says: 'Let us write every permutation of the 26 letters of the French alphabet, taken two at a time, ordering them alphabetically, and then do the same for the permutations of three letters, ... Everything that could be written is found in the list whose rule of formation we have just indicated.'

Therefore, the class of phrases which could be written in any language (using a finite number of words) is denumerable, or has the power of N_1. Then the class of ideas that can be expressed by ordinary language is denumerable, and so, also, is the class of decimal numbers (proper fractions) that can be defined in ordinary language (and by the signs of arithmetic and of algebra, which we can state in ordinary language).

The author defines:

$E =$ 'decimal numbers that can be defined in ordinary language,'

and constructs with the denumerable E the number which occurs in Cantor's theorem of the preceding section. That is, if fn is the number of class E at the nth place, the author considers the number

$$N = \Sigma[10^{-n} \text{ anti Cfr}_{-n}fn|n, N_1]$$

and lets G be the phrase which translates the second member into words of ordinary language. Then:

The number N does not belong to the class E, for reasons explained in the preceding section; and it does belong to the class E, because it is defined by a finite number of words in the language considered. But this is a contradiction.

The author resolves the contradiction as follows: 'The phrase G of letters is one of the permutations, and so will be in my list. But there it has no meaning. It is the set E which is in question, and that has not yet been defined.'

This could be reworded as follows: 'The error in the preceding contradiction is in the second affirmation, for E is a class of numbers defined by words of ordinary language, whereas N is a number defined by words of ordinary language *and by the letter E*, which has no meaning in ordinary language. Therefore we may not conclude that N belongs to the class E.'

But the class E is defined by words of ordinary language. Therefore, if we substitute for E its definition, the result will be that N is expressed by words of ordinary language alone, and the antinomy remains.

The number N was defined in symbols by proposition (1).

The defining member contains the constant signs Cfr, anti, Σ, ..., and the variable sign f. The letter n is an apparent variable. Therefore N is defined by fn, 'the element in the nth place of class E.'

We continue the transformation of the definition into symbols.

Let the letters $a, b, c, ...$ be digits in some system of enumeration. The base B of the system is the number of them, say 25. If we add other typographical signs for 'point,' 'space,' ..., the number B is still finite. I shall give the name 'alphabetic system' to the system of enumeration to base B, where the digits have the form $a, b, c, ..., z, 0$. This last cannot be an initial sign (e.g., the typographical period).

Then the numbers N_1 are expressed by

$a, b, ..., z, a0, aa, ab, ..., zz, a00, a0a, ...,$ one, ..., six, ..., two, ..., one divided by three, ...

Every sequence of letters is a natural number N_1, written in the alphabetic system.

If n is a natural number (N_1), and if it, written in the alphabetic system, determines a sequence of letters that has a meaning in the language considered, and defines a decimal number (Θ), we define:

Value n = the decimal number, which the number n, written in the
    alphabetic system, defines, according to the rules of ordinary
    language.   (2)

'Value n' is not defined for every number n.

The natural numbers which, when expressed in the alphabetic system, determine a phrase which represents a decimal, constitute the class

$N_1 \cap x \ni (\text{Value } x \, \varepsilon \, \Theta)$.

The nth number of this class is represented, according to the conventions of the Formulario v (p. 120), by

$\min_n[N_1 \cap x \ni (\text{Value } x \, \varepsilon \, \Theta)]$,

and the 'nth element of E' is expressed by

$fn = \text{Value } \min_n[N_1 \cap x \ni (\text{Value } x \, \varepsilon \, \Theta)]$.   (3)

Note that fn is the value of the nth phrase which expresses a decimal. Several phrases may express the same number. The question is not essentially varied if we consider only decimals different from the preceding ones.

Formulas (1), (2), (3) express the definition of N.

The sign E is not necessary. The whole definition of N is made up of proposition (1), which defines N by f; (3), which defines f by 'Value'; and (2), which defines 'Value.' Propositions (1) and (3) are in symbols. The definition of 'Value' is expressed in ordinary language, and may contain hidden difficulties.

4 'THE TRUE SOLUTION' OF MR POINCARÉ

Poincaré gives the following solution (p. 307) of the antinomies of Richard and others: 'E is the set of *all* the numbers that can be defined by a finite number of words, *without introducing the notion of the set E itself.* Without that the definition of E would contain a vicious circle.'

If by 'notion of E' we understand the sign E, then the affirmation of the author agrees with common usage, and is part of the rules given in the Formulario v, p. 14, for the definition of any sign x is an equality of the form

$x = $ (expression composed of signs preceding x).

The preceding rule eliminates, in an easy and mechanical way, every possibility of a vicious circle.

The rule of Poincaré does not eliminate the possibility of a vicious circle, for it excludes x from the second member, but says nothing about excluding the signs that come after x.

Now, the definitions of Richard do not contain a vicious circle. This results from the definitions (1) and (3) of §3, expressed in symbols, and (2), expressed in ordinary language, the weak point in the argument.

And so, the solution of the antinomy given by Poincaré, which accuses the definitions of a vicious circle, is not exact.

If by 'notion of E' we understand an expression equivalent to E, then in every definition the second member, or defining member, being equivalent to the first, or defined member, always contains the notion of the first, so that Mr Poincaré contradicts the rules of common usage and renders impossible any definition.

For example, the definition of difference given in every book (Formulario v, p. 44) is

$$a \, \varepsilon \, N_0 \, . \, b \, \varepsilon \, a + N_0 \, . \, \supset \, . \, b - a = \imath \, N_0 \, \cap \, x \, \mathfrak{z} \, (a + x = b) \qquad \text{Def.}$$

'If a is a number, and b is a number $\geqslant a$, then $b - a$ indicates that number x which satisfies the condition $a + x = b$.'

Mr Poincaré, page 315, says (the italics indicate words of the author; the rest relates to my example):[3] '*The fault is again the same*; $b - a$ *is* among *all* numbers that which satisfies the condition; *under penalty of a vicious circle, this must mean* among *all the* numbers *in the definition of which the notion of* minus *does not enter. This excludes* the number $b - a$, *which depends on* minus. *The definition of $b - a$, then, is not predicative.*'

The definition of root can only be given in the form

$$a \, \varepsilon \, Q \, . \, \supset \, . \, \sqrt{a} = \imath \, Q \, \cap \, x \, \mathfrak{z} \, (x^2 = a) \qquad \text{Def.}$$

3 [The original of this article was written in *Latino sine flexione*. Peano kept the words of Poincaré in the original French; they are distinguished here by being italicized.]

'If a is a positive real number, then $\sqrt{a}$ indicates that positive real number which has a as its square.'

Poincaré would say: 'The fault is again the same; $\sqrt{a}$ is among all numbers that which has a as its square; under penalty of a vicious circle, this must mean among all the numbers in the definition of which the notion of $\sqrt{a}$ does not enter. This excludes $\sqrt{a}$.'

On page 316, Poincaré says: 'The word *all* has a well-defined meaning when it is a question of a finite number of objects. In order for it to have meaning when the objects are infinite in number, it would be necessary to have an actual infinity. Otherwise *all* the objects could not be conceived as having been posited before their definition, so that if the definition of a notion N depends on *all* objects A, a vicious circle cannot be avoided if among the objects A there is one which cannot be defined without using the notion N itself.'

For example, consider the definition of least common multiple, as it has been given by every author from Euclid to our day (*Formulario*, v, p. 53):

$$a, b \, \varepsilon \, N_1 \, . \, \supset \, . \, m(a, b) = \min[(a \times N_1) \cap (b \times N_1)] \qquad \text{Def.}$$

'If a and b are natural numbers, then m(a, b), read (following Lebesgue, Lucas, ...) the least common multiple of a and b, indicates the minimum among the multiples of a and b.'

Poincaré would say: 'The word multiples of a and b has no meaning, for it constitutes an actual infinity. If the definition of m(a, b) depends on all objects $a \times N_1 \cap b \times N_1$, there is a vicious circle if among these objects there is one, m(a, b), which cannot be defined without m(a, b) or an equivalent phrase.'

I have referred above to three passages of Poincaré where he expresses his ideas on definitions. But the objection to Poincaré's solution is so obvious that I doubt whether I understand the author well. Let me give another example.

In analysis it is known that

$$x \, \varepsilon \, q \, . \, \supset \, . \, e^{ix} = \cos x + i \sin x \qquad (1)$$

$$e^{ix} = \text{known series} \qquad (2)$$

$$\cos x, \sin x = \text{known series.} \qquad (3)$$

Euler's formula (1) expresses the exponential by trigonometric functions, and expresses both the sine and cosine functions by the exponential. Therefore, exp contains the notions of sin and cos; sin and cos contain the notion of exp.

The rules of mathematical logic say: the ordering of the exponential and trigonometric functions is arbitrary. If we presuppose that exp has been defined by (2), then sin and cos are rigorously defined by (1).

If we presuppose that sin and cos have been defined by (3), then (1) is a legitimate definition of the exponential. This entirely conforms to the universal opinion. The rule of Poincaré renders illegitimate every definition of form (1).

In conclusion, the Formulario, in the part on mathematical logic, contains either explicitly or by citation the rules of definition and proof in mathematics which are found in writers from Aristotle and Leibniz to the present day. The rules for definitions and proofs, gathered together in the Formulario by the collaboration of many, are compared with the theories of various authors and are applied to an enormous number of definitions and proofs in mathematics. In general these rules are satisfied by authors of mathematics. If a rule is not satisfied, the defect in the definition and proof is indicated, as in the case of the principle of Zermelo, known now for fourteen years.

Poincaré is reconstructing mathematical logic on his own, independently of all previous studies. His rules for definitions are either too broad, and hence do not eliminate defects, or they destroy the whole of mathematics. The genius of Poincaré did not arrive at the simple rule

$x =$ expression composed of preceding signs,

a necessary and sufficient rule for the elimination of vicious circles.

5 A NEW SOLUTION

We now proceed to the calculation of

$N =$ that decimal number which has as its nth decimal digit the antidigit of the nth digit of the decimal number expressed by the nth phrase among the phrases expressing numbers in ordinary language, ordered alphabetically,

as it was defined by Richard, after the substitution for E of its definiens.

Let us imagine a sequence of phrases, or of natural numbers written in an alphabetic system. The first phrases which express decimals are (let us suppose)

one divided by six $= 0.1666 \ldots,$
one divided by two $= 0.5,$
two divided by six $= 0.333 \ldots;$

then $N = 0.658 \ldots$

The sequence of phrases with 15 letters, like the preceding italicized phrases, would form a line much longer than the distance from the earth to the sun. We continue on to longer phrases, to the definitions of π, of 'e,' etc. When we treat phrases of 100 letters, the line of digits calculated for N surpasses every measure of the physical world. We continue the calculation, and reach phrases of 200 letters where, say, we finally reach the phrase defining N.

Let $m - 1$ be the number of digits already calculated for N when the phrase defining N occurs. We must now calculate the mth digit of N. But the mth phrase among the phrases expressing numbers is the phrase defining N, so that the rule says:

'The mth digit of N is equal to its own antidigit,'

or $\mathrm{Cfr}_{-m} N = \mathrm{anti}\ \mathrm{Cfr}_{-m} N$

or $x - = x$,

which is absurd.

Therefore, one of the conditions that determine the sequence of digits of N is contradictory. The phrase defining N only appears to define a number. In ordinary language we say, 'the number N does not exist'; in symbols, we may not write

$N \, \varepsilon \, \Theta$.

The preceding definition is similar to the following:

$N = $ (largest prime number) (Euclid)
$N = $ (the number which satisfies the condition x not $= x$) (Richard)
$N = $ (the real number which satisfies the condition $x^2 + 1 = 0$)
$N = $ (the real number which satisfies the condition $x^2 - 1 = 0$)
$N = $ (lim sin x, for $x = \infty$)
$N = $ (derivative of $|x|$, for $x = 0$)

which have the appearance of numbers, but do not indicate numbers. The reason is obvious, and is stated in the Formulario.

In the preceding examples the word 'the' occurs, which corresponds to the sign $\imath$ of the *Formulario*. In the words 'largest,' 'lim,' and 'derivative,' the sign $\imath$ is in the definition. See the Formulario v for the definitions of 'largest' (p. 46) and of 'limit' (p. 214). 'Derivative' is defined by 'limit.'

In every case, in order that we may deduce from the definition some proposition of the form

$\imath \, a \, \varepsilon \, u$, 'the a which is a u,'   '$\imath \, a$ is in the class u,' it is necessary and sufficient that the conditions which are given in the Formulario, in the definition of ι (p. 13), be satisfied:

$$a \, \varepsilon \, \text{Cls} \, . \, \exists \, a : x, y \, \varepsilon \, a \, . \supset_{x,y} . \, x = y$$

'a is a non-empty class, and if we suppose two individuals x, y are in the class a, then $x = y$,' or 'a is a condition which determines one and only one individual.'

In the examples cited, the class which follows 'the' is either empty (as in Richard's case) or contains several individuals.

Is this to be the final solution? There remains a weak point in the definition of N, its expression in ordinary language. If the phrase defining N does not express a number, as I have shown above, then in calculating N I go beyond that phrase, since it does not define a number, and the definition of N makes sense. That is, if N does not exist, then it exists.

This new antinomy is similar to the affirmations:

What I am saying is false,
What I am saying is nonsense,

which if they are true are false, and if they are false are true.[4] The contradiction is in the ambiguity of the phrase N. It is necessary to explicitly add 'this phrase included' or 'this phrase excluded.'

Then let us not consider the ambiguous phrase N, and proceed further. A bit further along we find the phrases:

$N' = $ (phrase N), this phrase excluded;
$N'' = $ (phrase N), this phrase included.

N'' has no value, for the reasons given. N' represents a determinate number, belonging to the class E and differing from every *other* number E, as is clear.

The weak point, however, in the remarkable example of Richard is that the definition of N is given partly in symbols (Prop. 1, 3, pp. 150, 151) and partly not in symbols (Prop. 2). The non-symbolic part contains ideas of 'ordinary language,' ideas very familiar to us, but indeterminate and the cause of every ambiguity. The example of Richard does not belong to mathematics, but to linguistics; one element, fundamental to the definition of N, cannot be defined exactly (according to the rules of mathematics). From elements which are not well defined we can deduce several mutually contradictory conclusions.

4 Russell, *The Principles of Mathematics, passim.*

XIX
On the foundations of analysis (1910)*

When the idea of publishing English translations of some articles of Peano was proposed to Ugo Cassina, editor of the Opere scelte *of Peano, he especially recommended the inclusion of the following selection. It is the text of a talk given by Peano in 1910 to the Piedmont section of the 'Mathesis' Society, which had been founded in 1895 by teachers of mathematics in the middle schools. In it we find Peano's mature views on the teaching of mathematics. Here is that clarion call for rigour and simplicity that is still relevant today. Here is the linguistics scholar using his skill to place the problem linguistically as well as historically. Here is Peano, the teacher!*

Students begin their study of mathematics in elementary schools. In secondary schools, they begin afresh with definitions. In universities, the various courses in analysis begin with definitions of the various kinds of numbers. Calculus texts, which began at one time with derivatives, now feel the need to begin with the theory of limits, followed by the theory of irrationals, and then by definitions of the other kinds of numbers. Our graduates, however, realize that they still have confused ideas about the foundations of mathematics and, if they are in a position to teach others, eagerly look for books on the subject, now that they are free to study what they choose. Many such books are published each year. They shed new light on controversial questions, but sometimes obscure ideas which were formerly thought to be clear. The following example is an illustration of this.

Legendre introduced the following well-known definition of a straight line in place of Euclid's definition, which he thought to be less clear: 'A straight line is the shortest distance between two points.' This definition appeared very clear in the time of Legendre, and will appear so to pupils just beginning the study of geometry. Everyone knows, however, that the idea of *distance* or *length* of a line, which appears in this definition, is much more complicated than the idea, straight line, which is to be defined: calculus

* 'Sui fondamenti dell'Analisi,' 'Mathesis' società italiana di matematica, Bollettino, 2 (1910), 31–7 [143].

texts are still not agreed on a definition of the length of a line, or on the relative theory. Thus, the definition of Legendre, which expresses a simple idea by means of one more complicated, must be rejected, as, indeed, modern texts have done.

The study of the foundations of mathematics having become ever broader and more interesting, societies have been founded with the purpose of pursuing and promoting it. Everywhere there are societies whose purpose is to make their members rich in gold (love of which is the cause of so much evil). There are societies which defend the material and moral interests of their members. Professors of mathematics, however, have founded societies in France, Germany, England, and America, whose purpose is to treat the scientific, philosophic, and didactic problems which touch on their teaching. Italy, too, and with the same pure, noble, and disinterested purpose, has seen the rise of the *Mathesis* Society, one of the first in time, but, unfortunately, one of the last in material means.

The study of these philosophico-didactic questions is, above all, satisfying to the human mind in its continual search for truth. It is, indeed, interesting to discover, in a path worn by generations down through the centuries, new topics of study and new theories which require all the keenness of our intellect.

Furthermore, this study is, in essence, of immediate usefulness – on a level with a discovery which would allow us to travel faster, or would lower the price of bread – because a knowledge of these questions, and how to solve them, has the effect of perfecting our teaching, of speeding up the advance of our students, and of giving them the necessary fund of information at the cost of less effort.

Persons who prefer the more material pleasures to the pleasures of philosophy have a prejudicial disregard for these studies. Such questions, they say, can never be treated in the secondary schools; for that reason, all discussion and study of them is useless. It is true that these questions cannot be treated in school, but it is necessary that the teacher know the solution, or solutions, of them, so that he may know how to select the best, and not repeat just that one which he studied in school; and it is essential that he know the questions that do not have a solution, on which he has to be silent. A person who does not know well the foundations of any part whatever of mathematics will always remain hesitant, with an exaggerated fear of rigour.

Others believe that mathematics is, at least in its foundations, fixed and occupied always with repeating that two and two make twenty-two. When they hear talk of new studies and of doubts about the methods that they are used to following, they take it as a personal insult and, on every occasion, show their dislike of rigour – sometimes to the point of becoming dangerous. *Cave canem!*

Mathematical rigour is very simple. It consists in affirming true statements and in not affirming what we know is not true. It does not consist in affirming every truth possible. Science, or truth, is infinite: we know only a finite part of it, a part infinitely small with respect to the whole. Of the science which we do know, we have the obligation of teaching only that part which is most useful to our students.

In order to be rigorous, therefore, it is not necessary to define all the entities which we are considering. In the first place, everything cannot be defined. The initial entity cannot be defined, as Aristotle observed, and definition itself cannot be defined, i.e., the sign $=$. The most modern studies reduce to around ten the number of entities pertaining to the various branches of mathemetics, viz., logic, arithmetic, and geometry, which we do not know how to define.

Furthermore, even where a definition can be made, it is not always useful to do so. Every definition expresses a truth, namely that an equality can be established whose first member is the entity being defined and whose second, or defining member, is an expression composed of entities already considered. Now, if this truth has a complicated form, or is in some way difficult to explain, it belongs to that order of truths which are not suitable to be taught in a given school, and which may be passed over in silence.

Thus, it is known today that all geometrical entities may be defined with the aid of two: *point* and *distance* between two points. Before Professor Pieri, in 1898, and Professor Hilbert, in 1899, brought it to light, however, this truth was unknown. Even now, when we are in possession of it, there is still not a single school text based on this reduction. There is, then, a response to the question of geometrical definitions in the scientific field, but not yet in the didactic.

Likewise, every proof expresses a truth of another order, that a given proposition is a logical consequence of preceding propositions. If a proof is complicated, however, we may omit this truth. For example, the proof of the commutative, associative, and distributive properties of addition and multiplication, i.e., the reduction of these to others more simple, was begun by Leibniz, who was the first to show that $2 + 2 = 4$. He was followed by Grassmann, and this analysis has been continued by Dedekind and others. Nevertheless, everyone always knew these truths: we still find illiterate farmers using them in carrying out by memory multiplications they need, attaining a height of mental calculation which is miraculous to people used to doing calculations on paper. For this reason, the proof of these truths constitutes a new truth whose importance may well be less than that of other theories. Such proofs could well be omitted.

In consequence, any book whatever, even one full of blunders, may be made rigorous by leaving out what is false; what remains is the useful part of the book. We see, then, that rigour produces simplicity and economy in teaching.

Therefore, from many arithmetic texts still infesting our schools, let us take out the definition: 'Addition is that operation by which two or more numbers are summed (united),' which expresses *to add* by means of the synonym *to sum* or *to unite*. Let us suppress the definition of *to subtract*, which uses *to remove* or *to take away*, words themselves not defined. These pseudo-definitions are based on the fact that ordinary language has a multitude of different forms to express the same idea, a richness – or waste – of words. These sentences, which are passed off as definitions, give no idea of addition or subtraction, and if the pupils learn to understand and carry out these operations, in spite of these definitions, they would succeed even better if no definitions at all were given – which, indeed, is already the practice in some schools.

Let us suppress from so-called Euclidean geometries the definitions of *point* (using *part*), *line* (using *length*), *surface* (using *width*), *straight line*, and *plane*, which all express the unknown by the unknown.[1] Let us suppress all this and put nothing in its place, because points, straight lines, and planes are always there before our eyes.

We may suppress from the arithmetic of Euclid the definition (Book VII, 1) Μονάς ἐστιν, καθ' ἣν ἕκαστον τῶν ὄντων ἕν λέγεται, which substantially means: '*Unity* is that which is one.' We may do the same with the definition of *number*.

For centuries, operations have been performed with negative numbers, fractions, irrationals, and imaginaries; it is only the latest generation which has been occupied with giving them rigorous definitions. This work, however, is not finished: the theories being expounded do not all agree and often new theories are presented. For example, the definition of fractions, already treated by so many in various ways, was again made the object of a competition by your worthy *Mathesis* Society during the congress last autumn in Padua. It is notable that the members would deprive themselves of a part of their far from sumptuous salaries in order to hear a talk on fractions, and it is even more notable that several, urged not by the prize, but by a common ideal, were induced to make this an object of study and

---

1 It is curious to observe that Aristotle, a hundred years before Euclid, reported and criticized several definitions of Euclid. Thus the definition γραμμὴ ἐστι μῆκος ἀπλατες (a line is length without width) was criticized because it is negative.

a subject of their lessons. The prize was won by Professor Padoa, of Genoa, with his new theory.

My criticism of the work of Euclid or of others is not meant to show a lack of respect, but rather is a measure of my esteem. The work of Euclid challenges the centuries. In vain the great Legendre attempted, a century ago, a geometry superior to Euclid's; in vain in the past fifty years have new texts in the geometry of Kant's theories of space been introduced: the book of Euclid always appears superior to the rest. Moreover, a definition is indicated in Greek by ὁρισμός, and Aristotle always speaks thus. The propositions of Euclid, criticized as definitions, carry, instead, the title ὅροι – and ὅρος means *term*. Hence, these pages are a collection of terms, like a dictionary, and the relative propositions are not definitions, or postulates, or theorems – they are clarifications. The reading of ὅρος as *definition* is not appropriate.

In general, criticism must always be understood in the etymological and benevolent meaning of the word. Only one who has worked can be subjected to criticism and discussion. He who does not work is apart from all criticism.

At any rate, definitions of unknowns by unknowns still abound in mathematics. For several years I taught: 'We call *series* a *succession* of quantities formed by a determined *law*,' and it was only later that I realized that the words *succession* and *law* would have to be defined in turn.[2]

If we suppress from an arithmetic text whatever is inexact or superfluous, such as the multiplicity of equivalent terms, the book will be greatly reduced, and will end by being reduced to symbols alone. For several years (more precisely, from around 1889), by a purely theoretical route, I and others have come to writing in symbols for the purpose of analysing some difficult questions of mathematics – especially relative to the foundations. But others before us, led by purely didactic reasons, arrived at the same result. To speak only of our country, Professor Gerbaldi, now at the University of Pavia, published an arithmetic for use in the elementary school. I have a copy of the 1888 edition, which is entirely in symbols. This arithmetic, published at the expense of the city of Rome on the advice of the late Professor Cerruti, was used for many years with great success. It represents absolute rigour.

Symbols are all that may be imagined to be most rigorous and, at the same time, simple, clear, and easy for the pupils. The new methods are difficult, not for the pupils, but for the teachers. They are obliged to make an effort to study them, and to look, from a new point of view, at things which were formerly presented to them under a different aspect. We are all

2 The same applies to the definition of *function* using *correspondence*, of *collection* using *set* and *property*, etc.

occupied with our own work, and unable to follow up every novelty. Some, on seeing such novelties, can correctly say: 'These are indeed beautiful, but I am unable to study them.' On the other hand, others pretend to have infused knowledge: they say *a priori* that the book is difficult, that they do not understand it, and how then can they make the pupils understand it? Then they expose to ridicule the pupils in the secondary schools who do understand.

To recognize whether a theory is exact, natural logic is needed. Its methods, which have been studied and classified, constitute – to the extent they refer to mathematics – mathematical logic. An external criterion consists in consulting several books on the same question. If we see differences of opinion, we may in general conclude that the theory is not yet consolidated.

One of the most controversial points (I limit myself to this) is the introduction of the various kinds of numbers – natural, negative, fractional, irrational, and imaginary. Here the sequence ends, seeing that the definitions of the various other numbers – complex, quaternions, etc. – do not present further difficulties.

The difficulty in the definition of the various entities is in part linguistic. The word *number* (*natural number*) having been introduced as a translation of the word ἀριθμός of Euclid, the phrase *prime number* indicates both in grammar and in arithmetic a class of numbers, just as white man indicates a class of men. On the other hand, the phrase *whole number* does not indicate a class of numbers (as it grammatically should), but a larger class than that of the numbers. Here the adjective does not restrict the class to which it is applied, but enlarges it. Likewise, *rational number* indicates a class, not contained in, but containing the second, and *real number* indicates a fourth class larger than the third. This nomenclature, contrary to common usage, is not found in Euclid, although some very interesting calculations with irrationals are developed there. It is quite recent. In consequence, it was wished by all means to have the phrase 'whole number' indicate a number, and so the 'principle of permanence' was made by Hankel in 1867.

Let us see this principle in action. One usually begins by premissing certain propositions whose validity is in no way doubtful. These are:

There does not exist any number (from the sequence 0, 1, ...), which when added to 1 gives 0 as a result.

There does not exist a number (integral), which multiplied by 2 gives 1.

There does not exist a number (rational), whose square is 2.

There does not exist a number (real), whose square is -1.

Then one says: in order to overcome such an inconvenience, we extend the concept of number, that is, we introduce, manufacture, create (as

Dedekind says) a new entity, a new number, a sign, a sign-complex, etc., which we denote by -1, or $\frac{1}{2}$, or $\sqrt{2}$, or $\sqrt{(-1)}$, which satisfies the conditions imposed. In other words:

$$-1 = \text{that number } x \text{ such that } x + 1 = 0,$$
$$\tfrac{1}{2} = \qquad\qquad " \qquad\qquad x \times 2 = 1,$$
$$\sqrt{2} = \qquad\qquad " \qquad\qquad x^2 = 2,$$
$$\sqrt{(-1)} = \qquad\qquad " \qquad\qquad x^2 = -1.$$

If, in the second member, we understand by *number* that which until that instant had that name, then the entities considered are self-contradictory: hence, several names have been given to a non-being. Or if, in the second member, a new entity is understood by *number*, then we have the unknown defined by the unknown. To say that this is 'entirely different from all numbers' is to say that which is not, and not that which is. It is then natural to ask oneself why it is that we create new entities here, and not in other impossible cases. There does not exist a largest prime number: to make arithmetic more general, let us manufacture an ideal prime number, larger than all the prime numbers. Two parallel lines have no common Euclidean point: we imagine a point at infinity. Two oblique lines have no common point: to take away all exceptions, we attribute to them a transfinite common point. 'Can anyone keep from laughing?' asked Gauss, apropos this introduction of imaginaries. 'This would be playing with words, or rather misapplying them.'

This principle of permanence reached its apogee with Schubert, who, in the *Encyclopädie der mathematischen Wissenschaften*, affirmed that one must 'prove that for numbers in the broad sense, the same theorems hold as for numbers in the narrow sense.' Now, if all the propositions which are valid for the entities of one category are valid also for those of a second, then the two categories are identical. Hence – if this could be proved – the fractional numbers are integers! In the French edition of the *Encyclopédie* these things are put to rights. There it says that one must be 'guided by a concern for keeping the formal laws as much as possible.' Thus the principle of permanence acquires the value of a principle, not of logic, but of practice, and is of the greatest importance in the selection of notation. Basing their work on precisely this principle – a particular case of what Mach called the principle of economy of thought – Professors Burali-Forti and Marcolongo succeeded in untangling the disordered skein of notations in vectorial calculus, where all used to be arbitrary (and many still believe that the notations are necessarily arbitrary).

The entities we first considered, -1, $\frac{1}{2}$, $\sqrt{2}$, $\sqrt{(-1)}$, may be defined by looking for them in a known category that contains them. They come about by abstraction, or as operations.

The questions I have touched on are all the more interesting, the more they are studied. Anyone who has time may consult Russell's masterpiece, *The Principles of Mathematics* (1903). Other, more recent books are those of Mannoury and of Klein. Poincaré, too, has begun to occupy himself with these matters. The *Periodico di matematica* of Professor Lazzeri often contains interesting articles on these subjects. The latest number contains a very nice article of Professor Cipolla on fractions. Worthwhile too, from the same viewpoint, are many articles in the *Bollettino di matematica* published by Professor Conti. These studies, besides being of scientific interest, immediately make the teacher more fit for his position, and this is a benefit which is passed on and generates other benefits.

I wish the Mathesis Society complete success for its excellent purpose, for the merit of its members, and, in a special way, for the untiring efforts of its founder.

Finally, I thank those here now who had the courage to sacrifice their day of rest in order to listen to chit-chat about the fundamentals of analysis, instead of a presentation of them.

XX
The importance of symbols in mathematics (1915)*

In the following note Peano briefly traces the history of mathematical symbols, and discusses their importance and the reasons for it. He closes with a discussion of the importance of symbols in mathematical logic, to the development of which he had contributed. But that was already at an earlier time, and Peano had been criticized for not keeping abreast of the developments. His answer here is that he was interested in mathematical logic only for its use in mathematics, and not for its philosophical interest. The answer is resigned, but not bitter, and he returns a compliment by praising the work of Bertrand Russell. Peano would have agreed with Norbert Wiener's decision in 1913 to spend his travelling fellowship in England because he had 'learned that Peano's best days were over, and that Cambridge was the most suitable place for a training in mathematical logic.'

The splendid and highly imaginative article of Eugenio Rignano, 'Le forme superiori del ragionamento,' *Scientia*, 17 (1915), 11–37, has persuaded me to treat an analogous question, namely the function that symbols have in mathematics.

The oldest symbols, which are also the most widely used today, are the digits used in arithmetic, 0, 1, 2, etc., which we learned about 1200 from the Arabs, and they from the Indians, who were using them about the year 400.

The first advantage that one sees in the digits is their brevity. Numbers written with Indo-Arabic digits are much shorter than the same numbers spelled out in ordinary language, and are even shorter than the same numbers written with the Roman numerals I, X, C, M.

Further reflection reveals that these symbols are not just shorthand, i.e., abbreviations of ordinary language, but constitute a new classification of ideas. Thus even though the digits 1, 2, ..., 9 correspond to the words 'one, two, ..., nine,' the words 'ten, hundred' have no corresponding simple symbols, but rather the composite symbols '10, 100.' Furthermore, the symbol 0 has no equivalent in our native language; we use 'zero,' a con-

* 'Importanza dei simboli in matematica,' *Scientia*, 18 (1915), 165–73 [176].

traction of the Arabic word, while the Germans and the Russians use the Latin word 'null.' Symbolism does not consist in the form of the symbols: Europeans have been using a fixed form since the invention of printing, which is quite different from that used by the Arabs. Nevertheless our digits are correctly called Indo-Arabic because they have the same value as the corresponding Arabic digits.

The use of digits not only makes our expressions shorter, but makes arithmetical calculation essentially easier, and hence makes certain tasks possible, and certain results obtainable, which could not otherwise be the case in practice.

For example, direct measure assigned to the number π, the ratio of the circumference of a circle to its diameter, the value 3. The Bible tells us that Solomon constructed a vase of bronze ten cubits in diameter and thirty cubits in circumference (I Kings 7:23), so that $\pi = 3$.

Archimedes, about 200 B.C., by inscribing and circumscribing polygons about a circle, or rather by calculating a sequence of square roots, using Greek digits, found π to within 1/500. The substitution of Indian digits for the Greek allowed Aryabhata, about the year 500, to extend the calculation to 4 decimal places, and allowed the European mathematicians of 1600 to carry the calculation out to 15 and then 32 places, still following Archimedes' method. Further progress, i.e., the calculation of 100 digits in 1700, and the modern calculation of 700, was due to the introduction of series.

The same thing may be said for the symbols of algebra, $+$, $-$, $\times$, $=$, $>$, in universal use today. Algebraic equations are much shorter than their expression in ordinary language, are simpler and clearer, and may be used in calculations. This is because algebraic symbols represent ideas and not words. For example, the symbol $+$ is read 'plus,' but the symbol and the word do not have the same value. We say 'a is b plus a quantity' and we write '$a > b$,' without 'plus' corresponding to the sign $+$, and the expression 'sum of a and b' is translated into symbols as '$a + b$,' although the phrase does not contain the word 'plus.' The symbol $+$ allows us to represent that which in ordinary language is expressed by 'plus, sum,' and even 'addition, term, polynomial.' Likewise, the sign $\times$ represents, without being equivalent to, the words 'multiplication, product, factor, coefficient.' Algebraic symbols are much less numerous than the words they allow us to represent.

We could not conceive of an algebra today without symbols, but in fact all the propositions of algebra now studied in middle school are found in Euclid and Diophantus without symbols. There ordinary words have assumed a special technical meaning, having already assumed the value of phonetic symbols. Thus the Euclidean phrase 'the ratio of the number a to the number b' is exactly expressed by our symbol a/b, the word 'ratio,' or λόγος in Euclid, having only a remote common origin with the commonly used word, and with the word 'logic' which is derived from it.

The evolution of algebraic symbolism went like this: first, ordinary language; then, in Euclid, a technical language in which a one-to-one correspondence between ideas and words was established; and then the abbreviation of the words of the technical language, beginning about 1500 and done in various ways by different people, until finally one system of notation, that used by Newton, prevailed over the others.

The use of algebraic symbolism permits students in middle school easily to solve problems which previously only great minds like Euclid and Diophantus could solve, and permits the treatment of many new algebraic questions.

The symbolism of infinitesimal calculus is a continuation of that of algebra. Here the history is more certain. Archimedes measured the area of several figures by using a form of reasoning called 'method of exhaustion.' Kepler in 1605, Cavalieri in 1639, Wallis in 1665, and others said that the area described by the ordinates of a curve is the *sum* of all the ordinates. Leibniz abbreviated the word 'sum' by the initial S. Bernoulli called this an 'integral' and it now has the form of an elongated S.

The expression of an unknown area as the sum of an infinite number of ordinates, an undefined sum, would seem to be the expression of the obscure by the more obscure, but in fact this sum, or integral, has some of the fundamental properties of ordinary sums and this makes calculation easy. These ordinates, whose sum is the area, are the 'indivisibles' of Cavalieri and the 'infinitesimals' of Leibniz. Most geometers of the period rejected these methods, saying that one could only find known results with them (and this was true up to a certain point), that all the results obtained could also be found with the older methods, and they proved this by redoing the proofs in the language of Archimedes (which is always possible). Then they grew tired of doing it, and everyone adopted the new symbolism, which is much easier to use.

Geometry has made less use of symbolism. Analytic geometry gives an algebraic appearance to geometrical questions. It is a method of study, powerful at times, but indirect and often inferior to elementary geometry. Many were looking for a calculus that would operate directly on the geometrical entities. Hérigone in 1644, Carnot in 1801, and many others used symbols for 'line, plane, parallel, perpendicular, triangle, square, etc.'

Some of these are used in modern texts in elementary geometry, but they are merely shorthand symbols and do not lend themselves to any calculus.

The modern theory of vectors allows the treatment of geometrical questions with a direct calculus similar to algebra. The germ of the idea of a vector is found in Euclid, more clearly in several authors about 1800, and in our Bellavitis in 1832. He called them 'segments,' because a segment in elementary geometry is very like a vector, but this use of an old word with a new meaning caused confusion, all the more serious just because the two entities were so similar. Hamilton in 1845 called this new entity a 'vector' and in a two-volume work explained its very simple calculus, applying it to questions of elementary geometry, analytic geometry, projective and infinitesimal geometry, as well as to astronomy and mathematical physics, giving a new and simpler form to known results and finding new ones. But the studies of Hamilton did not reach the continent until Maxwell, who was also English, adopted this method for the presentation of his theory of electricity and magnetism.

Today vectors are known, at least by name, by all mathematicians and are used by many, but up to now they have not been included in the official school program. Now it happened that several authors took the liberty of changing notations, of introducing new and unnecessary symbols, and of giving the name of vector to similar, but not identical, entities, bringing back the ambiguity that Hamilton had eliminated. I am not speaking of those who attribute contradictory properties to vector, making any calculus impossible. From that came such confusion that the illustrious mathematician Laisant recently posed, in the periodical *L'Enseignement mathématique* of Geneva, the question: *What is a vector*?

This tangled skein was unravelled by Professors Burali-Forti and Marcolongo in several articles published in the *Rendiconti* of Palermo. It results that the notation cannot be entirely arbitrary but must follow definite laws. These professors, in collaboration with Professors Boggio, Bottasso, and others, have begun the publication of a series of volumes treating the principal applications of vectors, so that the best books on this theory, which used to be printed in England, are now published in Italy.

What is a vector? A vector is not a segment to which other properties are added. A vector cannot be defined with elementary geometry, i.e., one cannot write an equality in which the first member is the word 'vector' and the second member is a group of words from elementary geometry. A vector results from a segment, or rather from a pair of points, by abstracting from several of its properties. The equality of vectors can be defined.

If A, B, C, D are points, the expression $A - B = C - D$ means that the segments AB and CD are of equal length, and have the same direction.

That is, *A*, *B*, *D*, *C* are consecutive vertices of a parallelogram or, in other words, the mid-point of *AD* coincides with the mid-point of *BC*. The vector *A* − *B* is the set of all properties common to the vectors *C* − *D* which are equal to *A* − *B*. As a consequence we may speak of the sum of two vectors, but not of the origin of a vector, of the line which contains it, of adjacent vectors, etc., because substituting one vector for its equal can change its origin.

The theory of vectors presupposes no knowledge of analytic geometry or even of elementary geometry. Professor Andreoni is using it in the industrial school in Reggio Calabria to explain geometry and it could very well be used in middle school.

The symbolism of mathematical logic, or logical calculus, or algebra of logic, was the last to appear, but already in its present development it shows itself to be in no way inferior to the preceding symbolism of arithmetic, algebra, and geometry.

In every book of mathematics there are terms, or symbols, which represent ideas of algebra or of geometry. The remaining terms, about a thousand, represent ideas of logic. Mathematical logic classifies the ideas of logic presented in mathematical books, represents them by symbols, and studies their properties, or the rules of logical calculus. In this way the whole book is expressed in the symbols of mathematics and of logic.[1]

The first advantage seen in the symbols of logic is the brevity which they produce. Thus my *Formulario* contains a complete treatment of arithmetic, of algebra, of geometry, and of infinitesimal calculus, along with definitions, theorems, and proofs, all in a small volume, much smaller than the volumes that contain the same things expressed in ordinary language.

Next we see that, whereas the words of ordinary language which express logical relations are about a thousand, the symbols of mathematical logic which express the same ideas are ten, just as many as the Arabic numerals. And this is just in practice, seeing that Professor Padoa, in his book already

1 The authors who use the symbols of logic usually explain them in the first page of their works. The *Rivista di matematica*, edited by me, volume 7, pages 3–5, contains a list of 67 works from 1889 to 1900 relating to mathematical logic; and the *Formulario mathematico*, published by me, 5th edition, pages xiv–xv, contains a list of 62 works from 1900 to 1908. Others have appeared afterwards. Among them the *Algebra der Logik* of Schröder merits special mention. It contains a very rich bibliography, especially of older works.

The reader who wishes to have a more ample knowledge of this subject may consult C. Burali-Forti, *Logica matematica* (Milan: Hoepli, 1894) and the more recent book, up to date with the latest results, A. Padoa, *La logique deductive dans sa dernière phase de développement*, extracted from the *Revue de métaphysique et de morale* (Paris, 1912).

cited, has reduced the number of symbols theoretically necessary to only three. It thus results that the ideographic symbols are much less numerous than the words that they can express, and hence there is no one-to-one correspondence between symbols and words. These symbols are not abbreviations of words, but represent ideas.

The principal utility of the symbols of logic, however, is that they make reasoning easier. All those who use logical symbolism attest to this.

Mario Pieri, taken from science in the flower of his activity in 1913, adopted the symbols of logic in a series of memoirs relative to the foundations of projective geometry which were published from 1895 on. Of the greatest importance is his work 'Della geometria elementare come sistema ipotetico deduttivo' [*Mem. Accad. sci. Torino* (2), 49 (1899)]. The first definitions met in the usual texts in geometry, the definitions of point, of line, of plane, etc., do not satisfy a logician. To say that 'a line is length without width' is to express the unknown idea *line* by the two even more unknown ideas *length* and *width*. Pieri analysed and classified the ideas of geometry and expressed all of them as functions of two primitive ideas: *point* and *distance* between two points.[2]

The largest work entirely written in ideographic symbols is A.N. Whitehead and B. Russell, *Principia mathematica*.[3] The authors, in their preface,

2　The results reached by Pieri constitute an epoch in the study of the foundations of geometry. B. Russell, of Cambridge University, in his book *The Principles of Mathematics* (1903, p. 382), says of the work of Pieri: 'This is, in my opinion, the best work on the present subject.' And all those who later treated the foundations of geometry have made ample use of Pieri's work, and have rendered judgments equivalent to that given by Russell. I may cite Bôcher, in the *Bulletin of the American Mathematical Society* (1904, p. 115); Wilson, in the same periodical (1904, p. 74); Huntington, in the *Transactions* of the same society, in a series of memoirs from 1902 on; Veblen in the same periodical (1904); and many others.

Pieri, in other works, presented his results without using symbols, but he always insisted that he found them by using the symbols of mathematical logic. See, for example, his communication to the International Congress of Philosophy (Paris, 1900) with the title, 'La Géométrie comme système purement logique,' especially page 381. Then, too, having been asked to give the inaugural address for the year 1906–7 at the University of Catania, Pieri selected as his topic 'Uno sguardo al nuovo indirizzo logico-matematico delle scienze deduttive' ['A look at the new logicomathematical tendency in the deductive sciences']. This address was printed in the *Yearbook* of that university and is a most clear presentation of this great scientific movement, stated so as to be accessible to the non-mathematical public, which could quickly form from it a clear idea of this subject.

At that period of time a number of illustrious Italian mathematicians worked in the same direction. Thus, in 1900, L. Couturat, while stating that 'the Italian school has obtained results marvelous for rigour and subtlety,' was still uncertain 'whether one should attribute this to the use of logical symbolism or to the insight of the scholars who used it.' But in 1905 he unhesitatingly affirms that 'it is the indispensable tool for attaining logical purity of concepts and deductive rigour in reasoning.'

3　Cambridge, University Press, vol. I, 1910, 666 pp.; vol. II, 1912, 722 pp.; vol. III, 1913, 491 pp.

explain the utility, indeed the necessity, of symbolism. They say that they were obliged to use symbols rather than words because the ideas used in their book are more abstract than those considered in ordinary language, and hence there are no words having the exact value of the symbols. Indeed the abstract and simple ideas considered in their work lack an expression in ordinary language, which more easily represents complex ideas. Symbolism is clearer for this reason. It also allows the construction of a series of arguments in which the imagination would be entirely unable to sustain itself without the aid of symbols. This work treats the foundations of analysis and of geometry, the theory of point sets, the infinite, the infinitesimal, limits, and all the most difficult and controversial questions of mathematics.[4]

Mathematical logic is not only useful in mathematical reasoning (and this is the only use I have made of it), but is also of interest in philosophy. Louis Couturat, accidentally killed at the beginning of the war of 1914, wrote numerous important articles in the *Revue de métaphysique et de morale*, as well as books and monographs.

It has been recognized that several forms of syllogisms considered in scholastic logic lack some condition, that the rules given for definitions in treatises of scholastic logic do not apply to mathematical definitions, and that vice versa the latter satisfy other rules which are not found in the usual texts in logic. The same is true of the rules for mathematical proofs, as they cannot be reduced to the syllogism of classical logic, but rather assume other, completely classified, forms.

---

4 Among the authors of other important works in mathematical logic we may limit ourselves to mentioning the following: Professor Huntington of Harvard University in a series of papers published in the *Transactions of the American Mathematical Society*, beginning in 1902, analysed the ideas of magnitude, real number, groups of substitutions, etc., using logical symbolism in part, and stated that he made use of the works of Professors Burali-Forti, Padoa, Couturat, and Amodeo. Professor Moore of the University of Chicago applied the symbolism of mathematical logic in his study of the new problem of integro-differential equations in a communication to the Fourth International Congress of Mathematicians (Rome, 1908), and then in his book *Introduction to a Form of General Analysis* (1910). The same method was applied by Doctor Maria Gramegna, a victim of the earthquake in Avezzano in January of this year, in the paper 'Serie di equazioni differenziali lineari,' *Atti Accad. Sci. Torino*, 45 (1909–10), 469–91.

I should also mention the works of Professor Cipolla of the University of Catania, relative to congruences, published in the *Rivista di matematica*. Also, the well-received book: G. Pagliero, *Applicationes de calculo infinitesimale* (Turin: Paravia, 1907), entirely written in symbols. In the educational field may be mentioned the various texts of arithmetic and algebra of Professor Catania, of Catania, in which symbols are not used but the results of mathematical logic are applied, and thus a clear, simple, and rigorous presentation results, qualities which are seldom found together. Finally, it is owing to the collaboration of Professors Castellano, Vacca, Vailati, and others, that the *Formulario mathematico* has reached its present state.

It is necessary then to distinguish one work from another in 'mathematical logic.' If Eugenio Rignano has justly criticized those who consider mathematical logic as a science in itself (and it is very true that their works are often of little use), on the other hand it would not be just to criticize those (whose works have been cited by me) who consider mathematical logic as a useful tool for resolving mathematical questions which resist the usual methods. For that matter, Rignano himself recognizes this when he affirms the usefulness of our *Formulario* and that it completely attains its goal. That from such a new symbolic tool new results can be obtained can be seen from the general opinion of those who have used it. That these new results are important can be seen from the fact that the works written in logical symbols have been read and cited by numerous authors, and have served as a basis for new research.

XXI
Definitions in mathematics (1921)*

Peano more than once concerned himself with the question of definitions in mathematics. Between 1911 and 1915 he published six articles having 'definition' in the title. Our final selection dates from 1921 and is his third article with the same title 'Definitions in mathematics.' This question continued to occupy him and in the fall of 1931 he assigned the study of it to Ludovico Geymonat as a minor thesis topic for the degree in mathematics.

Aristotle gave several rules for definitions and proofs in general, and these rules have been reproduced in treatises of scholastic logic almost unaltered to our day. Lately, however, people devoting themselves to mathematical, or symbolic, logic, on re-examining this question, have found that a number of scholastic rules do not apply to mathematical definitions, and that these definitions satisfy rules not previously stated. Let us examine successively several rules.[1]

1 EVERY DEFINITION IS AN EQUALITY

Every definition has the form:

the defined = the thing defining,

where the defined is a new sign or word or sentence or proposition, and the defining is an expression composed of known signs. The definition expresses the convention of using the defined in place of the longer defining phrase. This is clear in the equalities

$$2 = 1 + 1, \qquad 3 = 2 + 1, \qquad 4 = 3 + 1, \qquad \text{etc.,}$$

---

* 'Le definizioni in matematica,' *Periodico di matematiche* (4), 1 (1921), 175–89 [193].
1 Several of the observations which follow have already been published by me in an article with the same title in French, in *Congrès international de philosophie* [Paris, 1900]; then translated into Polish and republished in Warsaw in 1902; then in an expanded article in the *Instituto de ciencias de Barcelona* in 1911. Many other authors treated the same question then and later; they will be cited subsequently.

which can be considered as definitions of the digits 2, 3, 4, etc., supposing known the digit 1, and the sign of addition, even if only in the case in which the second term of the sum is 1, i.e., the operation $+1$.

The common definitions of the transcendental numbers,

$$e = \lim (1 + 1/m)^m, \quad \text{for } m \text{ infinite,}$$

and Euler's constant,

$$\gamma = \lim (1 + 1/2 + 1/3 + \ldots + 1/n - \log n), \quad \text{for } n \text{ infinite,}$$

are also equalities.

If the ideographic sign $=$ is not written between the two members of the definition, it is expressed by terms of ordinary language, for which it may be substituted.

For example, Euclid's first definition, 'a point is that which has no parts,' may be written:

point $=$ 'that which has no parts.'

His second definition may be written:

line $=$ 'length without width.'

The defined may be composed of several words. Such are the common definitions of 'straight line,' of 'prime number,' etc.

The defined may also be a proposition or a relation. Of this type is the definition of parallelism between two lines, in Euclid (I, 23): 'Two lines are mutually parallel' $=$ 'They lie in the same plane and have no common point no matter how far extended.'

In the same way another of Euclid's definitions (VII, 13) may be written: 'Two numbers are mutually prime' $=$ 'They have no common divisor except unity.'

Hence, to reduce every definition to an equality, it is necessary to use the sign $=$, not only between two numbers, but also between two classes, between two propositions, and between every kind of entity.

A proposition of the form

$$\pi = 3.1415 \ldots$$

cannot be a definition, even though it is presented as an equality. It is an improper equality, indicating only the first digits of π, or rather it signifies in a precise way that

$$3.1415 < \pi < 3.1416,$$

which is not an equality, and hence cannot be a definition.

Euclid's third definition, 'the extremes of a line are points,' cannot be a definition. It is not the definition of *point* (defined by proposition 1 above); it is not the definition of *line* (defined by 2); it is not the definition of *extreme*, because the affirmation that the extremes of a line are points expresses a property of extremes, which is not enough to distinguish them. Proposition 3 just given has the name *definition* in the version of Euclid published by Heiberg (Leipzig, 1883) and in almost all versions, but Euclid's text gives to all the propositions of that chapter the generic name ὅροι, and the Greek ὅρος means *term*.

This may be seen in Definition 13, 'term, ὅρος, is the extreme of something,' and from Definition 8 of Book v, where ὅρος indicates 'term of a proportion.' Correctly, then, Professor Vacca, in his version, says: 'I translate the Greek ὅροι by *terms*, rather than by *definitions*, as is usually done, because these first introductory pages, instead of being mathematical definitions, are rather clarifications, or explanations, analogous to those given today in dictionaries.'[2]

Besides, it should be remembered that the books of Aristotle and Euclid, and somewhat later the grammar of Donatus, the arithmetic of Boethius, and all the scholastic books, have been transmitted to us by means of successive copies made by teachers and students; and every copyist added to or modified the text as he pleased, so that it is difficult to recognize what parts are due to which authors.

The rule that every definition be an equality is found implicitly in Aristotle, although the word *equality* is not found there. See A. Pastore, 'Le definizioni matematiche secondo Aristotele e la logica matematica,' *Atti Accad. sci. Torino*, 47 (1911–12), 478–94. The author, a professor of theoretical philosophy at the University of Turin, compares there scholastic logic and mathematical logic.

2 ALL DEFINITIONS IN MATHEMATICS ARE NOMINAL

The custom in scholastic logic is to classify definitions as real and nominal. In mathematics all definitions are nominal. This is well known. Pascal, *Pensées*: 'We recognize in geometry only those definitions which logicians call nominal definitions.' Möbius (*Werke*, I, p. 388) in 1815: 'The division of definitions into verbal and real has no meaning.'

2 G. Vacca, *Euclide, il primo libro degli elementi, testo greco, versione italiana, introduzione e note*, Florence, 1916, price two lire. I warmly recommend this book to teachers in middle schools. Whoever has retained some knowledge of Greek can read the father of geometry in the original, and whoever does not know Greek, by using the little Greek-Italian vocabulary included, will be able to read the original without having to spend years studying grammar. Several young university graduates succeeded in a short time in reading the Greek in this book.

In general for all definitions, Mill (1838): 'All definitions are of names, and names only.' Definitions in natural history, which are said by some to be 'real,' are said by others, and with greater reason, to be 'descriptions' of animals or of plants.

3 THE RULE OF GENUS AND DIFFERENCE DOES NOT HOLD FOR ALL DEFINITIONS

Aristotle (*Top.*, ɪ, 8) sets the rule:

ὁ ὁρισμὸς ἐκ γένους καὶ διαφορῶν ἐστίν,

which Boethius renders 'by proximate genus and specific difference,' and this is reproduced as an absolute rule in all treatises of logic. The classic example of this property is the definition

man = rational animal.

Here 'animal' and 'rational' indicate two classes. Between these two classes is understood the operation called conjunction by the grammarians, logical multiplication by the logicians following Boole, indicated in general by *and* in ordinary language and in mathematical logic by the sign ∩.

Hence the rule of Aristotle would say that every definition has the form

$$x = a \cap b,$$

where a and b are known classes, called genus and species, and x is the class which they define.

Some mathematical definitions satisfy the Aristotelian rule. Such is Euclid's Definition 22, which may be translated:

square = quadrilateral ∩ equilateral ∩ equiangular.

But this rule may be applied only to definitions of a class. It is not true for the definition $2 = 1 + 1$, for that of the number e given above, or for definitions of entities which are not classes. Even definitions of classes do not necessarily have the preceding form. For example, in the definition

composite number = (number greater than 1) × (number greater than 1),

between the two (in this case, identical) classes the sign of logical conjunction ∩ is not used, but rather the sign of arithmetical multiplication.

The proponents of classical logic answer that in the definitions '2 = sum of 1 with 1,' 'composite number = product of two numbers,' 'e = limit of etc.,' the genus is represented by the words *sum, product, limit*. These words, however, do not indicate classes, but rather functions; every number is a sum and a product and a limit. *Class* corresponds to the first category

($o\dot{v}\sigma\acute{\iota}a$) of Aristotle, while *function* pertains to the fourth ($\pi\rho\acute{o}s$ $\tau\iota$), and here there is need to distinguish these two concepts. A definition of class using *collection* or *set* or *property* is a vicious circle, as is that of *function* using *relation* or *correspondence* or *operation*. Having arrived at these elementary ideas, we cannot proceed except with symbols. See, for example: G. Vailati, 'Aggiunte alle note storiche del Formulario,' *Rivista di matematica*, 8 (1902–6), 57–8; G. Vailati, 'La teoria Aristotelica della definizione,' *Rivista di filosofia e scienze affini* (1903), reprinted in *Scritti di G. Vailati*, pp. 485–96; L. Couturat, *Les principes des mathématiques* (Paris, 1905).

4 THE EXISTENCE OF THE THING DEFINED IS NOT NECESSARY

Some logicians affirm that one must define only existing things. Among them is Mill, who, starting from the definition of something non-existent and supposing it existent, arrived at an absurd result. But the absurdity derives from supposing existent that which was defined, not from having defined something non-existent.

The example of Mill, transformed into geometry, is the following:

From the proposition

(1) pentahedrons are polyhedrons

is deduced

(2) regular pentahedrons are polyhedrons,

from which, converting with the rule 'if every *a* is a *b*, then some *b* is an *a*,' we have

(3) some polyhedron is a regular pentahedron,

which is not so. Mill holds proposition (2) to be false, whereas mathematical logic holds (2) to be correct, but considers illegitimate the passage from (2) to (3). It is not true that the universal proposition 'every *a* is a *b*' converts immediately to the particular 'some *b* is an *a*,' but rather 'if every *a* is a *b*, and some *a* exist, then some *b* is an *a*.' Consequently all the forms of the syllogism in which a particular consequence follows from universal premises are incomplete; the conclusion follows from the stated hypotheses and from an existential hypothesis, left tacitly understood in scholastic logic, but explicitly stated in mathematical logic. See, for example, A. Padoa, *La logique déductive dans sa dernière phase de développement* (Paris, 1912). Aristotle (*Anal. post.*, Book I, Ch. 10, n. 9) also says: 'Definitions are not hypotheses, since they do not say that things exist or do not exist.'

In mathematics definitions of non-existent things are numerous. Euclid

(Book IV, prop. 20) gives a name to the largest prime number and, reasoning on it, concludes that it does not exist; i.e., he defines a name with the purpose of proving that it does not represent anything.

Neither the definition of limit, nor that of derivative, nor that of integral of a function affirms the existence of the entity defined. The definition of integral is always followed by a series of propositions of the form:

> 'If the function is continuous, the integral exists.'
> 'If the function is increasing, the integral exists.'
> 'A necessary and sufficient condition for the existence of the integral is that ...'

From the ordinary definition, 'the derivative of a function is the limit of its incremental ratio,' it results that the derivative exists or not according to whether this limit exists or not. Some authors, wishing to be more rigorous, say, 'the derivative is the limit, *where it exists*, of the incremental ratio,' and then, if the limit does not exist, we cannot conclude any longer that the derivative does not exist.

Moreover, the word *exist* has, in ordinary language, several meanings. The null class represents a class in which no individuals exist; but *it* exists. Thus the number 0 may indicate either the absence of size or the size 0. In practice it is necessary to define important things, whether existing or not.

5 POSSIBLE DEFINITIONS

We shall call 'possible definition' every equality which contains in one member a sign which does not appear in the other. For example, possible definitions of the number π are the following:

π = circumference : diameter.

π = minimum positive root of the equation
$$x - x^3/3! + x^5/5! - \dots = 0.$$

$$\pi = \int_{-\infty}^{+\infty} \frac{dx}{1 + x^2}.$$

The first is the historical definition, and is intelligible to the reader who knows the elements of geometry. The second is purely analytic. The third requires integral calculus.

Let us imagine the ideas we are considering to be disposed in a certain order. We shall call 'possible definition relative to a given order' an equality which contains in the first member a sign, and in the second an expression composed of signs preceding it. For example, the three possible definitions given for π are respectively possible to someone knowing elementary geometry, or the theory of series, or integral calculus.

Among the various possible definitions of an entity, in a fixed order, the choice of the actual definition depends on the will of the author and is made for didactic reasons. Hence a proposition may or may not be a possible definition – that depends on its nature. That a proposition is a definition depends on the will of the author, and not only on the nature of the proposition.

6 THE DEFINITION MUST PROCEED FROM THE KNOWN TO THE UNKNOWN

This is an evident rule. We find in Aristotle (*Top*, VI, ch. IV, n. 2) that one must see to it that 'definitions are made using things which are previous and better known': εἰ μὴ διὰ προτέρων καὶ γνωριμωτέρων πεποίηται τὸν ὁρισμόν.

Some treatises of mathematics define each new word by means of those studied in the preceding pages. This is not the case, however, for the first definitions, and these authors are not accustomed to saying which ideas they are supposing known. If they do not tell us, it is impossible to list them. Consider for example the first definitions of Euclid-Legendre:

'A line is a length without width.'
'A surface is that which has length and width, without height or depth.'

We see defined here the words 'line, surface' by using the not previously defined geometrical ideas 'length, width, height, depth.' The ideas which appear in the second members are more numerous than those in the first. It is therefore natural to ask whether it is not convenient to suppress these definitions, assuming as undefined ideas line and surface. Indeed, the idea of length, which is here taken as intuitive, is defined later in treatises of calculus, and the definition is not simple.

Some authors presuppose known ordinary language. Belonging to ordinary language, however, are the terms 'point, line, plane, sphere, one, two, three,' etc. For this reason the analysis of the question – whether the simplest elements of geometry, ideas of number, and the simplest ideas of arithmetic can be defined or not – requires an analysis of the language, the listing of all words and grammatical flexions which are presented in the first pages of these sciences, and hence the reconstruction of an ideography.

Ordinary language has numerous synonyms; whoever defines one of these using another, and so on, neither defines nor analyses the idea, but makes a vicious circle, called a 'circular definition.'

The words *add, sum, adjoin, unite* are synonyms, and a vicious circle is made in defining one of these words by using another. The same happens when one defines *subtract* by *take away, multiply* by *times, divide* by *partition*. Circular also are the definitions of *series* by *sequence, class* by *aggregate*, and *proposition* by *judgment*.

In the phrases 'five fingers,' 'five metres,' 'five hundred,' the apposition of the two words has the value of *multiplied*, and it is impossible to define the apposition using more appositions. The commonest definition of multiplication, 'the product of two numbers is the sum of as many terms equal to the multiplicand as there are units in the multiplier,' is a sonorous phrase, but it cannot give the idea to anyone who does not have it already. See, for example, R. Frisone, 'Le varie definizioni di prodotto,' *Atti Accad. sci. Torino*, 53 (1917–18), 420–7. There the author shows in which point of the sentence is found the apposition which has the same meaning as that which is to be defined, and examines various definitions of product. See also the same author's 'Le prime definizioni in aritmetica,' *Bollettino di matematiche e scienze fisiche e naturali*, 18 (1916–17), 80–3.

A mathematical concept, which has been expressed here with an apposition, may be found elsewhere indicated by other grammatical forms.

Kepler (1605), Cavalieri (1639), and Wallis (1655) used the phrases 'the ordinates' and 'all the ordinates' to indicate what we call today 'integral of the ordinate';[3] i.e., the complex idea of integral was expressed by a plural ending, and was not defined there. Similarly in Euclid (also in Vacca's version) the plural indicates the sum.

7 DEFINITIONS BY INDUCTION

Students of mathematical logic classify mathematical definitions into several types.

We define by induction a function f of the numbers N_0 (the whole numbers starting from zero) by giving the value of $f(0)$ and then, whatever the number n, expressing $f(n + 1)$ using $f(n)$. Thus, supposing known $a + 1$, 'the successor of a,' we define the sum of two numbers

$$a + 0 = a,$$
$$a + (n + 1) = (a + n) + 1.$$

Definition of product:

$$a \times 0 = 0,$$
$$a \times (n + 1) = a \times n + a.$$

Definition of power:

$$a^0 = 1,$$
$$a^{n+1} = a^n \times a.$$

3 See the passages cited in *Formulario mathematico*, ed. v, pp. 352, 356, 359.

Definition of factorial:

$0! = 1,$

$(n + 1)! = n! \times (n + 1).$

These rigorous definitions give $a \times n$, a^n, $n!$, even for $n = 0$ and $n = 1$, cases which the ordinary definitions usually exclude, and they have been adopted in the school texts of Professor Catania.

8 DEFINITIONS BY ABSTRACTION

Sometimes we define a function $f(a)$ not by a nominal definition of the form $f(a) =$ expression composed of preceding signs, but by defining the equality $f(a) = f(b)$. For example, Definition 5 of Book 5 of Euclid is translated:

(the ratio of the size of a to the homogeneous entity $b =$ the ratio of c to d) = (for arbitrary natural numbers m and n, if ma is less than or equal to or greater than nb, then mc is less than or equal to or greater than nd).

This constitutes also the modern definition of real (irrational) numbers.

Examples from geometry:

(area of polygon $a =$ area of b) = (a and b can be decomposed into super-positional parts);

(the direction of line $a =$ direction of b) = (a and b are parallel).

Other examples may be found in my article 'Le definizioni per astrazione,' *Bollettino della 'Mathesis,'* 7 (1915), 106–20. See also the full and scholarly treatise: Burali-Forti, *Logica matematica*, 2nd ed. (Milan, 1919).

9 PRIMITIVE IDEAS

Even though an order has been given to the ideas of a science, not all of them can be defined. The first idea, since it has no predecessor, cannot be defined; the sign $=$, which figures in every definition, cannot be defined. We say that an idea is *primitive*, relative to a given order, if in this order of ideas we do not know how to define it. Therefore being a primitive idea is not an absolute characteristic, but is only such relative to the ideas which are supposed known.

The existence of primitive ideas is already found clearly expressed in Pascal: 'It is evident that the first terms one might wish to define presuppose preceding ones to be used in their explanation ... and so it is clear that we never arrive at the first. Thus, on pushing the search further and further we necessarily arrive at primitive terms, which cannot be defined.'

In the last thirty years the determination of the primitive ideas of mathematics has furnished occasion for numerous studies, almost all made with the aid of the ideography of mathematical logic. I cite in a special way the work of Burali-Forti, Padoa, and Pieri in Italy, Korselt in Germany, Gérard in France, Russell and Whitehead in England, Bôcher, Dickson, Huntington, and Veblen in America.

The late Professor Pieri, at that time professor at the University of Parma, in 1899 expressed all the ideas of geometry using only the two primitive ideas 'point' and 'distance between two points.'[4] Speaking of the work of Pieri, B. Russell, in his famous great book *The Principles of Mathematics*, says: 'This is, in my opinion, the best work on the present subject.'

The fundamental properties of the primitive ideas are determined by 'primitive propositions,' or propositions which are not proved, and from which all the other properties of the entities considered may be deduced. The primitive propositions function, in a certain way, as definitions of the primitive ideas. The authors cited have given us for various parts of mathematics several complete systems of primitive ideas, and of primitive propositions.

Aristotle (*Metaph.* 4, 4) stated the existence of primitive propositions: 'Not everything may be proved.' But my colleague Pastore and the late Vailati searched in vain in Aristotle for any hint of primitive ideas.

See the treatise: Shearman, *The Development of Symbolic Logic* (London, 1906).

10 HOMOGENEITY OF DEFINITIONS

If the definiendum depends on any variables, it is necessary that the definiens contain the same variables, and no others; i.e., the definiendum and the definiens must be homogeneous in their variables.

We may not define the resultant of two forces as that which has for components with respect to the coordinate axes the sums of the components of the given forces, because this defines the two forces not only by means of these forces but also by using the reference axes. The definition can be completed by adding that the result does not vary if the axes are changed.

If a, b, c, d are natural numbers, we have

$$a/b + c/d = (ad + bc)/(bd).$$

This expresses the common rule for the sum of the two fractions. But it may not be taken as a definition. Indeed, given the numbers a and b, the value

4 'Della geometria elementare, come sistema ipotetico deduttivo,' *Accad. sci. Torino Memorie*, (2) 49 (1899).

of the fraction a/b is determined; vice versa, given a fraction x, it may be put into an infinite number of forms a/b; or rather, although the value of the fraction is a function of the numerator and the denominator, on the other hand the numerator and denominator are not functions of the fraction, seeing that $1/2 = 2/4$, and the numerator of the first is not equal to the numerator of the second. In the case of the preceding formula, the first member is a function of the two variables a/b and c/d, whereas the second is a function of the four variables a, b, c, d. It is necessary to complete the definition by demonstrating that the second member is solely a function of a/b and of c/d.

To see better that it is not permissible to define an operation on fractions using one of their representations, let us define $x \mu y$, intermediate between the two fractions x and y, as the fraction which has for numerator the sum of the numerators and for denominator the sum of the denominators. That is, set

$$a, b, c, d \, \varepsilon \, \mathrm{N} \, . \, \supset \, . \, (a/b) \, \mu \, (c/d) = (a + c)/(b + d),$$

a formula of the same type as the preceding. We shall have $(1/2) \mu (2/3) = 3/5$, $(2/4) \mu (2/3) = 4/7$, and since $1/2 = 2/4$, we have $3/5 = 4/7$, which is a false result.

In mathematics there are many expressions which have the grammatical form of functions and are not functions. Besides *numerator* and *denominator* of a fraction, the words *terms* of a sum, *factors* of a product, *coefficient*, *base*, *exponent* are not functions of the value of the expression, but of its form. Equally the words *monomial, binomial, irreducible fraction* express properties of the form, not of the value of the expression. In an ideography, these words cannot be represented by symbols. Hence we see that the number of symbols of the *Formulario mathematico* is much less than the words of ordinary language.

11 UTILITY OF DEFINITIONS

Definitions are useful, but not necessary, since the definiens may always be substituted for the definiendum, thus eliminating the definiendum completely from the theory. This elimination is something very important. If by eliminating the defined symbol the new expression is not made longer and more complicated than the preceding, then this signifies that the definition was of little use. If difficulties are encountered in the elimination, this proves that the definition was not well devised. Indeed, this method of substitution is one of the best for recognizing the exactness of a definition. Aristotle (*Top.*, 6, 4) said: 'To make the inexactness of a definition become manifest, the concept must be put in place of the name.'

Rational numbers are defined using the natural numbers; hence every theorem about rationals may be transformed into a theorem about natural numbers only. This is easily done, and the language of Euclid is regained.

The irrational numbers are defined using the rational numbers; hence every theorem of analysis must ultimately be a theorem about natural numbers. This transformation has been attempted by some authors, but it is not easy. Among the results obtained, a very elegant one is that of R. Frisone, 'Una teoria simplice dei logaritmi,' *Atti Accad. sci. Torino*, 52 (1916–17), 846–53. The definitions and theorems on logarithms are explained there without speaking of irrationals, and hence only with integral exponents.

Theoretically definitions are arbitrary. Pascal said: 'Definitions are entirely free, and they are never subject to contradiction, for nothing is more permissible than to give a clearly designated thing any name we wish.' Leibniz, however, places a practical limitation: 'Definitions are arbitrary in themselves, but in practice they should conform to the common usage of one's colleagues.'

Hobbes (1642) had already said: 'If definitions are arbitrary, all mathematics, which is based on definitions, is an arbitrary science.' But supposing the definitions arbitrary, only the form of mathematics is arbitrary, not the content of the theorems. Even in the form we must follow the use of common mathematical language, abstaining from fabricating new names, or giving new meanings to known words, without necessity. If to a word of ordinary language is given a meaning very different, as is the case with *differential*, *integral*, *vector*, these become new words, and there is no danger of confusion.

If, however, a meaning which is only little different is given there is continual danger of confusion. For example, 'the value of π to four decimals,' according to the majority, is 3.1415; but others understand 3.1416; and others either of these two numbers. I could indicate treatises where the author, after having given one of these values as the definition, then uses it with the other meaning.

Mathematical logic, to which the present article pertains, has been the object of new studies. I shall mention the treatises of W.M. Kozlowski, *Podstawy logiki* (Warsaw, 1917), and C.I. Lewis, *A Survey of Symbolic Logic* (Berkeley, California, 1918), which is notable for its rich bibliography.

In conclusion, some mathematical definitions found in school texts are illusory. By simply suppressing them we gain in rigour and simplicity.

Index of names